Children's Books a

Illustrators

Gleeson White

Alpha Editions

This edition published in 2024

ISBN : 9789367242384

Design and Setting By
Alpha Editions
www.alphaedis.com
Email - info@alphaedis.com

As per information held with us this book is in Public Domain.
This book is a reproduction of an important historical work. Alpha Editions uses the best technology to reproduce historical work in the same manner it was first published to preserve its original nature. Any marks or number seen are left intentionally to preserve its true form.

THE INTERNATIONAL STUDIO
SPECIAL WINTER-NUMBER 1897-8

CHILDREN'S BOOKS AND THEIR ILLUSTRATORS. BY GLEESON WHITE.

There are some themes that by their very wealth of suggestion appal the most ready writer. The emotions which they arouse, the mass of pleasant anecdote they recall, the ghosts of far-off delights they summon, are either too obvious to be worth the trouble of description or too evanescent to be expressed in dull prose. Swift, we are told (perhaps a little too frequently), could write beautifully of a broomstick; which may strike a common person as a marvel of dexterity. After a while, the journalist is apt to find that it is the perfect theme which proves to be the hardest to treat adequately. Clothe a broomstick with fancies, even of the flimsiest tissue paper, and you get something more or less like a fairy-king's sceptre; but take the Pompadour's fan, or the haunting effect of twilight over the meadows, and all you can do in words seems but to hide its original beauties. We know that Mr. Austin Dobson was able to add graceful wreaths even to the fan of the Pompadour, and that another writer is able to impart to the misty twilight not only the eerie fantasies it shows the careless observer, but also a host of others that only a poet feels, and that only a poet knows how to prison within his cage of printed syllables. Indeed, of the theme of the present discourse has not the wonder-working Robert Louis Stevenson sung of "Picture Books in Winter" and "The Land of Story Books," so truly and clearly that it is dangerous for lesser folk to attempt essays in their praise? All that artists have done to amuse the august monarch "King Baby" (who, pictured by Mr. Robert Halls, is fitly enthroned here by way of frontispiece) during the playtime of his immaturity is too big a subject for our space, and can but be indicated in rough outline here.

THE "MONKEY-BOOK"   A FAVOURITE IN THE NURSERY
(By permission of James H. Stone, Esq., J.P.)

"ROBINSON CRUSOE." THE WRECK
FROM AN EIGHTEENTH-CENTURY CHAP-BOOK

Luckily, a serious study of the evolution of the child's book already exists. Since the bulk of this number was in type, I lighted by chance upon "The Child and his Book," by Mrs. E. M. Field, a most admirable volume which traces its subject from times before the Norman conquest to this century. Therein we find full accounts of MSS. designed for teaching purposes, of early printed manuals, and of the mass of literature intended to impress "the Fear of the Lord and of the Broomstick." Did space allow, the present chronicle might be enlivened with many an excerpt which she has culled from out-of-the-way sources. But the temptation to quote must be controlled. It is only fair to add that in that work there is a very excellent chapter to "Some Illustrators of Children's Books," although its main purpose is the text of the books. One branch has found its specialist and its exhaustive monograph, in Mr. Andrew Tuer's sumptuous volumes devoted to "The Horn Book."

"CRUSOE AND XURY ESCAPING"
FROM AN EIGHTEENTH-CENTURY CHAP-BOOK

Perhaps there is no pleasure the modern "grown-up" person envies the youngsters of the hour as he envies them the shoals of delightful books which publishers prepare for the Christmas tables of lucky children. If he be old enough to remember Mrs. Trimmer's "History of the Robins," "The

Fairchild Family," or that Poly-technically inspired romance, the "Swiss Family Robinson," he feels that a certain half-hearted approval of more dreary volumes is possibly due to the glamour which middle age casts upon the past. It is said that even Barbauld's "Evenings at Home" and "Sandford and Merton" (the anecdotes only, I imagine) have been found toothsome dainties by unjaded youthful appetites; but when he compares these with the books of the last twenty years, he wishes he could become a child again to enjoy their sweets to the full.

"CRUSOE SETS SAIL ON HIS EVENTFUL VOYAGE"
FROM AN EIGHTEENTH-CENTURY CHAP-BOOK

Now nine-tenths of this improvement is due to artist and publisher; although it is obvious that illustrations imply something to illustrate, and, as a rule (not by any means without exception), the better the text the better the pictures. Years before good picture-books there were good stories, and these, whether they be the classics of the nursery, the laureates of its rhyme, the unknown author of its sagas, the born story-tellers—whether they date from prehistoric cave-dwellers, or are of our own age, like Charles Kingsley or Lewis Carroll—supply the text to spur on the artist to his best achievements.

"THE TRUE TALE OF ROBIN HOOD."
FROM AN EIGHTEENTH-CENTURY CHAP-BOOK

It is mainly a labour of love to infuse pictures intended for childish eyes with qualities that pertain to art. We like to believe that Walter Crane, Caldecott, Kate Greenaway and the rest receive ample appreciation from the small people. That they do in some cases is certain; but it is also quite as evident that the veriest daub, if its subject be attractive, is enjoyed no less thoroughly. There are prigs of course, the children of the "prignorant," who babble of Botticelli, and profess to disdain any picture not conceived with "high art" mannerism. Yet even these will forget their pretence, and roar over a *Comic Cuts* found on the seat of a railway carriage, or stand delighted before some unspeakable poster of a melodrama. It is well to face the plain fact that the most popular illustrated books which please the children are not always those which satisfy the critical adult. As a rule it is the "grown-ups" who buy; therefore with no wish to be-little the advance in nursery taste, one must own that at present its improvement is chiefly owing to the active energies of those who give, and is only passively tolerated by those who accept. Children awaking to the marvel that recreates a familiar object by a few lines and blotches on a piece of paper, are not unduly exigent. Their own primitive diagrams, like a badly drawn Euclidean problem, satisfy their idea of studies from the life. Their schemes of colour are limited to harmonies in crimson lake, cobalt and gamboge, their skies are very blue, their grass arsenically green, and their perspective as erratic as that of the Chinese.

"TWO CHILDREN IN THE WOOD."
FROM AN EIGHTEENTH CENTURY CHAP-BOOK

"SIR RICHARD WHITTINGTON."
FROM AN EIGHTEENTH-CENTURY CHAP-BOOK

In fact, unpopular though it may be to project such a theory, one fancies that the real educational power of the picture-book is upon the elders, and thus, that it undoubtedly helps to raise the standard of domestic taste in art. But, on the other hand, whether his art is adequately appreciated or not, what an unprejudiced and wholly spontaneous acclaim awaits the artist who gives his best to the little ones! They do not place his work in portfolios or locked glass cases; they thumb it to death, surely the happiest of all fates for any printed book. To see his volumes worn out by too eager votaries; what could an author or artist wish for more? The extraordinary devotion to a volume of natural history, which after generations of use has become more like a mop-head than a book, may be seen in the reproduction of a "monkey-book" here illustrated; this curious result being caused by sheer affectionate thumbing of its leaves, until the dog-ears and rumpled pages turned the cube to a globular mass, since flattened by being packed away. So children love picture-books, not as bibliophiles would consider wisely, but too well.

"AN AMERICAN MAN AND WOMAN IN THEIR PROPER HABITS." ILLUSTRATION FROM "A MUSEUM FOR YOUNG GENTLEMEN AND LADIES" (S. CROWDER. 1790)

To delight one of the least of these, to add a new joy to the crowded miracles of childhood, were no less worth doing than to leave a Sistine Chapel to astound a somewhat bored procession of tourists, or to have written a classic that sells by thousands and is possessed unread by all save an infinitesimal percentage of its owners.

When Randolph Caldecott died, a minor poet, unconsciously paraphrasing Garrick's epitaph, wrote: "For loss of him the laughter of the children will grow less." I quote the line from memory, perhaps incorrectly; if so, its author will, I feel sure, forgive the unintentional mangling. Did the laughter of the children grow less? Happily one can be quite sure it did not. So long as any inept draughtsman can scrawl a few lines which they accept as a symbol of an engine, an elephant or a pussy cat, so long will the great army of invaders who are our predestined conquerors be content to laugh anew at the request of any one, be he good or mediocre, who caters for them.

It is a pleasant and yet a saddening thought to remember that we were once recruits of this omnipotent army that wins always our lands and our treasures. Now, when grown up, whether we are millionaires or paupers, they have taken fortress by fortress with the treasures therein, our picture-books of one sort are theirs, and one must yield presently to the babies as they grow up, even our criticism, for they will make their own standards of worth and unworthiness despite all our efforts to control their verdict.

If we are conscious of being "up-to-date" in 1900, we may be quite sure that by 1925 we shall be ousted by a newer generation, and by 2000 forgotten. Long before even that, the children we now try to amuse or to educate, to defend at all costs, or to pray for as we never prayed before—they will be the masters. It is, then, not an ignoble thing to do one's very best to give our coming rulers a taste of the kingdom of art, to let them unconsciously discover that there is something outside common facts, intangible and not to be reduced to any rule, which may be a lasting pleasure to those who care to study it.

It is evident, as one glances back over the centuries, that the child occupies a new place in the world to-day. Excepting possibly certain royal infants, we do not find that great artists of the past addressed themselves to children. Are there any children's books illustrated by Dürer, Burgmair, Altdorfer, Jost Amman, or the little masters of Germany? Among the Florentine woodcuts do we find any designed for children? Did Rembrandt etch for them, or Jacob Beham prepare plates for their amusement? So far as I have searched, no single instance has rewarded me. It is true that the *naïveté* of much early work tempts one to believe that it was designed for babies. But the context shows that it was the unlettered adult, not the juvenile, who was addressed. As the designs, obviously prepared for children, begin to appear, they are almost entirely educational and by no means the work of the best artists of the period. Even when they come to be numerous, their object is seldom to amuse; they are didactic, and as a rule convey solemn warnings. The idea of a draughtsman of note setting himself deliberately to please a child would have been inconceivable not so many years ago. To be seen and not heard was the utmost demanded of the little ones even as late as the beginning of this century, when illustrated books designed especially for their instruction were not infrequent.

"THE WALLS OF BABYLON." ILLUSTRATION FROM "A MUSEUM FOR YOUNG GENTLEMEN AND LADIES" (S. CROWDER. 1790)

As Mr. Theodore Watts-Dunton pointed out in his charming essay, "The New Hero," which appeared in the *English Illustrated Magazine* (Dec. 1883), the child was neglected even by the art of literature until Shakespeare furnished portraits at once vivid, engaging, and true in Arthur and in Mamillus. In the same essay he goes on to say of the child—the new hero:

"MERCURY AND THE WOODMAN." ILLUSTRATION FROM "BEWICK'S SELECT FABLES." BY THOMAS BEWICK (1784)

"And in art, painters and designers are vying with the poets and with each other in accommodating their work to his well-known matter-of-fact tastes and love of simple directness. Having discovered that the New Hero's ideal of pictorial representation is of that high dramatic and businesslike kind exemplified in the Bayeux tapestry, Mr. Caldecott, Mr. Walter Crane, Miss Kate Greenaway, Miss Dorothy Tennant, have each tried to surpass the other in appealing to the New Hero's love of real business in art—treating him, indeed, as though he were Hoteï, the Japanese god of enjoyment—giving him as much colour, as much dramatic action, and as little perspective as is possible to man's finite capacity in this line. Some generous art critics have even gone so far indeed as to credit an entire artistic movement, that of pre-Raphaelism, with a benevolent desire to accommodate art to the New Hero's peculiar ideas upon perspective. But this is a 'soft impeachment' born of that loving kindness for which art-critics have always been famous."

"THE BROTHER AND SISTER." ILLUSTRATION FROM
"BEWICK'S SELECT FABLES." BY THOMAS BEWICK (1784)

It would be out of place here to project any theory to account for this more recent homage paid to children, but it is quite certain that a similar number of THE STUDIO could scarce have been compiled a century ago, for there was practically no material for it. In fact the tastes of children as a factor to be considered in life are well-nigh as modern as steam or the electric light, and far less ancient than printing with movable types, which of itself seems the second great event in the history of humanity, the use of fire being the first.

"LITTLE ANTHONY." ILLUSTRATION FROM "THE LOOKING-GLASS OF THE MIND." BY THOMAS BEWICK (1792)

"LITTLE ADOLPHUS." ILLUSTRATION FROM "THE LOOKING-GLASS OF THE MIND." BY THOMAS BEWICK (1792)

To leave generalities and come to particulars, as we dip into the stores of earlier centuries the broadsheets reveal almost nothing *intended* for children—the many Robin Hood ballads, for example, are decidedly meant for grown-up people; and so in the eighteenth century we find its chap-books of "Guy, Earl of Warwick," "Sir Bevis, of Southampton," "Valentine and Orson," are still addressed to the adult; while it is more than doubtful whether even the earliest editions in chap-book form of "Tom Thumb," and "Whittington" and the rest, now the property of the nursery, were really published for little ones. That they were the "light reading" of adults, the equivalent of to-day's *Ally Sloper* or the penny dreadful, is much more probable. No doubt children who came across them had a surreptitious treat, even as urchins of both sexes now pounce with avidity upon stray copies of the ultra-popular and so-called comic papers. But you could not call *Ally Sloper*, that Punchinello of the Victorian era—who has received the honour of an elaborate article in the *Nineteenth Century*—a child's hero, nor is his humour of a sort always that childhood should understand—"Unsweetened Gin," the "Broker's Man," and similar subjects, for example. It is quite possible that respectable people did not care for their babies to read the chap-books of the eighteenth century any more than they like them now to study "halfpenny comics"; and that they were, in short, kitchen literature, and not infantile. Even if the intellectual standard of those days was on a par in both domains, it does not prove that the reading of the kitchen and nursery was interchangeable.

Before noticing any pictures in detail from old sources or new, it is well to explain that as a rule only those showing some attempt to adapt the drawing to a child's taste have been selected. Mere dull transcripts of facts please children no less; but here space forbids their inclusion. Otherwise nearly all modern illustration would come into our scope.

A search through the famous Roxburghe collection of broadsheets discovered nothing that could be fairly regarded as a child's publication. The chap-books of the eighteenth century have been adequately discussed in Mr. John Ashton's admirable monograph, and from them a few "cuts" are here reproduced. Of course, if one takes the standard of education of these days as the test, many of those curious publications would appear to be addressed to intelligence of the most juvenile sort. Yet the themes as a rule show unmistakably that children of a larger growth were catered for, as, for instance, "Joseph and his Brethren," "The Holy Disciple," "The Wandering Jew," and those earlier pamphlets which are reprints or new versions of books printed by Wynkyn de Worde, Pynson, and others of the late fifteenth and early sixteenth centuries.

Henry quitting School.

ILLUSTRATION FROM "SKETCHES OF JUVENILE
CHARACTERS"
(E. WALLIS. 1818)

In one, "The Witch of the Woodlands," appears a picture of little people dancing in a fairy ring, which might be supposed at first sight to be an illustration of a nursery tale, but the text describing a Witch's Sabbath, rapidly dispels the idea. Nor does a version of the popular Faust legend—"Dr. John Faustus"—appear to be edifying for young people. This and "Friar Bacon" are of the class which lingered the longest—the magical and oracular literature. Even to-day it is quite possible that dream-books and prophetical pamphlets enjoy a large sale; but a few years ago many were to be found in the catalogues of publishers who catered for the million. It is not very long ago that the Company of Stationers omitted hieroglyphics of coming events from its almanacs. Many fairy stories which to-day are repeated for the amusement of children were regarded as part of this literature—the traditional folk-lore which often enough survives many changes of the religious faith of a nation, and outlasts much civilisation. Others were originally political satires, or social pasquinades; indeed not a few nursery rhymes mask allusions to important historical incidents. The chap-book form of publication is well adapted for the preservation of half-discredited beliefs, of charms and prophecies, incantations and cures.

In "Valentine and Orson," of which a fragment is extant of a version printed by Wynkyn de Worde, we have unquestionably the real fairy story. This class of story, however, was not addressed directly to children until within the last hundred years. That many of the cuts used in these chap-books afterwards found their way into little coarsely printed duodecimos of eight or sixteen pages designed for children is no doubt a fact. Indeed the wanderings of these blocks, and the various uses to which they were applied, is far too vast a theme to touch upon here. For this peripatetic habit of old wood-cuts was not even confined to the land of their production; after doing duty in one country, they were ready for fresh service in another. Often in the chap-books we meet with the same block as an illustration of totally different scenes.

TITLE-PAGE OF "THE PATHS OF LEARNING" (HARRIS AND SON. 1820)

PAGE FROM "THE PATHS OF LEARNING" (HARRIS AND SON. 1820)

The cut for the title-page of Robin Hood is a fair example of its kind. The Norfolk gentleman's "Last Will and Testament" turns out to be a rambling rhymed version of the Two Children in the Wood. In the first of its illustrations we see the dying parents commending their babes to the cruel world. The next is a subject taken from these lines:

"Away then went these prity babes rejoycing at that tide,
Rejoycing with a merry mind they should on cock-horse ride."

And in the last, here reproduced, we see them when

"Their prity lips with blackberries were all besmeared and dyed,
And when they saw the darksome night, they sat them down and cried."

But here it is more probable that it was the tragedy which attracted readers, as the *Police News* attracts to-day, and that it became a child's favourite by the accident of the robins burying the babes.

The example from the "History of Sir Richard Whittington" needs no comment.

A very condensed version of "Robinson Crusoe" has blocks of distinct, if archaic, interest. The three here given show a certain sense of decorative treatment (probably the result of the artist's inability to be realistic), which is distinctly amusing. One might select hundreds of woodcuts of this type, but those here reproduced will serve as well as a thousand to indicate their general style.

Some few of these books have contributed to later nursery folk-lore, as, for example, the well known "Jack Horner," which is an extract from a coarse account of the adventures of a dwarf.

One quality that is shared by all these earlier pictures is their artlessness and often their absolute ugliness. Quaint is the highest adjective that fits them. In books of the later period not a few blocks of earlier date and of really fine design reappear; but in the chap-books quite 'prentice hands would seem to have been employed, and the result therefore is only interesting for its age and rarity. So far these pictures need no comment, they foreshadow nothing and are derived from nothing, so far as their design is concerned. Such interest as they have is quite unconcerned with art in any way; they are not even sufficiently misdirected to act as warnings, but are merely clumsy.

ILLUSTRATION FROM "GERMAN POPULAR STORIES." BY G. CRUIKSHANK (CHARLES TILT. 1824)

Children's books, as every collector knows, are among the most short-lived of all volumes. This is more especially true of those with illustrations, for their extra attractiveness serves but to degrade a comely book into a dog-eared and untidy thing, with leaves sere and yellow, and with no autumnal grace to mellow their decay. Long before this period, however, the nursery artist has marked them for his own, and with crimson lake and Prussian blue stained their pictures in all too permanent pigments, that in some cases resist every chemical the amateur applies with the vain hope of effacing the superfluous colour.

Of course the disappearance of the vast majority of books for children (dating from 1760 to 1830, and even later) is no loss to art, although among them are some few which are interesting as the 'prentice work of illustrators who became famous. But these are the exceptions. Thanks to the kindness of Mr. James Stone, of Birmingham, who has a large and most interesting collection of the most ephemeral of all sorts—the little penny and twopenny pamphlets—it has been possible to refer at first hand to hundreds, of them. Yet, despite their interest as curiosities, their art need not detain us here. The pictures are mostly trivial or dull, and look like the products of very poorly equipped draughtsmen and cheap engravers. Some, in pamphlet shape, contain nursery rhymes and little stories, others are devoted to the alphabet and arithmetic. Amongst them are many printed on card, shaped like the cover of a bank-book. These were called battledores, but as Mr. Tuer has dealt with this class in "The Horn Book" so thoroughly, it would be mere waste of time to discuss them here.

ILLUSTRATION FROM "GERMAN POPULAR STORIES." BY G. CRUIKSHANK (CHARLES TILT. 1824)

Mr. Elkin Mathews also permitted me to run through his interesting collection, and among them were many noted elsewhere in these pages, but the rest, so far as the pictures are concerned, do not call for detailed notice. They do, indeed, contain pictures of children—but mere "factual" scenes, as a rule—without any real fun or real imagination. Those who wish to look up early examples will find a large and entertaining variety among "The Pearson Collection" in the National Art Library at South Kensington Museum.

Turning to quite another class, we find "A Museum for Young Gentlemen and Ladies" (Collins: Salisbury), a typical volume of its kind. Its preface begins: "I am very much concerned when I see young gentlemen of fortune and quality so wholly set upon pleasure and diversions.... The greater part of our British youth lose their figure and grow out of fashion by the time they are twenty-five. As soon as the natural gaiety and amiableness of the young man wears off they have nothing left to recommend, but *lie by* the rest of their lives among the lumber and refuse of their species"—a promising start for a moral lecture, which goes on to implore those who are in the flower of their youth to "labour at those accomplishments which may set off their persons when their bloom is gone."

ILLUSTRATION FROM "THE LITTLE PRINCESS." BY J. C. HORSLEY, R.A. (JOSEPH CUNDALL. 1843)

The compensations for old age appear to be, according to this author, a little knowledge of grammar, history, astronomy, geography, weights and measures, the seven wonders of the world, burning mountains, and dying words of great men. But its delightful text must not detain us here. A series of "cuts" of national costumes with which it is embellished deserves to be described in detail. *An American Man and Woman in their proper habits*, reproduced on page 6, will give a better idea of their style than any words. The blocks evidently date many years earlier than the thirteenth edition here referred to, which is about 1790. Indeed, those of the Seven Wonders are distinctly interesting.

Here and there we meet with one interesting as art. "An Ancestral History of King Arthur" (H. Roberts, Blue Boar, Holborn, 1782), shown in the Pearson collection at South Kensington, has an admirable frontispiece; and one or two others would be worth reproduction did space permit.

ILLUSTRATION FROM "CHILD'S PLAY." BY E. V. B. (NOW PUBLISHED BY SAMPSON LOW)

Although the dates overlap, the next division of the subject may be taken as ranging from the publication of "Goody Two Shoes—otherwise called Mrs. Margaret Two-shoes"—to the "Bewick Books." Of the latter the most interesting is unquestionably "A Pretty Book of Pictures for Little Masters and Misses, or Tommy Trip's History of Beasts and Birds," with a familiar description of each in verse and prose, to which is prefixed "A History of Little Tom Trip himself, of his dog Towler, and of Coryleg the great giant," written for John Newbery, the philanthropic bookseller of St. Paul's Churchyard. "The fifteenth edition embellished with charming engravings upon wood, from the original blocks engraved by Thomas Bewick for T. Saint of Newcastle in 1779"—to quote the full title from the edition reprinted by Edwin Pearson in 1867. This edition contains a preface tracing the history of the blocks, which are said to be Bewick's first efforts to depict beasts and birds, undertaken at the request of the New castle printer, to illustrate a new edition of "Tommy Trip." As at this time copyright was unknown, and Newcastle or Glasgow pirated a London success (as New York did but lately), we must not be surprised to find that the text is said to be a reprint of a "Newbery" publication. But as Saint was called the Newbery of the North, possibly the Bewick edition was authorised. One or two of the rhymes which have been attributed to Oliver Goldsmith deserve quotation. Appended to a cut of *The Bison* we find the following delightful lines:

"The Bison, tho' neither
Engaging nor young,
Like a flatt'rer can lick you
To death with his tongue."

The astounding legend of the bison's long tongue, with which he captures a man who has ventured too close, is dilated upon in the accompanying prose. That Goldsmith used "teeth" when he meant "tusks" solely for the sake of rhyme is a depressing fact made clear by the next verse:

"The elephant with trunk and teeth
Threatens his foe with instant death,
And should these not his ends avail
His crushing feet will seldom fail."

Nor are the rhymes as they stand peculiarly happy; certainly in the following example it requires an effort to make "throw" and "now" pair off harmoniously.

"The fierce, fell tiger will, they say,
Seize any man that's in the way,
And o'er his back the victim throw,
As you your satchel may do now."

Yet one more deserves to be remembered if but for its decorative spelling:

"The cuccoo comes to chear the spring,
And early every morn does sing;
The nightingale, secure and snug,
The evening charms with Jug, jug, jug."

ILLUSTRATION FROM "THE HONEY STEW" BY HARRISON WEIR (JEREMIAH HOW. 1846)

But these doggerel rhymes are not quite representative of the book, as the well-known "Three children sliding on the ice upon a summer's day" appears herein. The "cuts" are distinctively notable, especially the Crocodile (which contradicts the letterpress, that says "it turns about with difficulty"), the Chameleon, the Bison, and the Tiger.

Bewick's "Select Fables of Æsop and others" (Newcastle: T. Saint, 1784) deserves fuller notice, but Æsop, though a not unpopular book for children, is hardly a children's book. With "The Looking Glass for the Mind" (1792) we have the adaptation of a popular French work, "L'Ami des Enfans" (1749), with cuts by Bewick, which, if not equal to his best, are more interesting from our point of view, as they are obviously designed for young people. The letterpress is full of "useful lessons for my youthful readers," with morals provokingly insisted upon.

"BLUE BEARD." ILLUSTRATION FROM "COMIC NURSERY TALES." BY A. CROWQUILL (G. ROUTLEDGE. 1845)

"Goody Two Shoes" was also published by Newbery of St. Paul's Churchyard—the pioneer of children's literature. His business—which afterwards became Messrs. Griffith and Farran—has been the subject of several monographs and magazine articles by Mr. Charles Welsh, a former partner of that firm. The two monographs were privately printed for issue to members of the Sette of Odde Volumes. The first of these is entitled "On some Books for Children of the last century, with a few words on the philanthropic publisher of St. Paul's Churchyard. A paper read at a meeting of the Sette of Odde Volumes, Friday, January 8, 1886." Herein we find a very sympathetic account of John Newbery and gossip of the clever and distinguished men who assisted him in the production of children's books, of which Charles Knight said, "There is nothing more remarkable in them than their originality. There have been attempts to imitate its simplicity, its homeliness; great authors have tried their hands at imitating its clever adaptation to the youthful intellect, but they have failed"—a verdict which, if true of authors when Charles Knight uttered it, is hardly true of the present time. After Goldsmith, Charles Lamb, to whom "Goody Two Shoes" is now attributed, was, perhaps, the most famous contributor to Newbery's publications; his "Beauty and the Beast" and "Prince Dorus" have been republished in facsimile lately by Messrs. Field and Tuer. From the *London Chronicle*, December 19 to January 1, 1765, Mr. Welsh reprinted the following advertisement:

"ROBINSON CRUSOE." ILLUSTRATION FROM "COMIC NURSERY TALES." BY A. CROWQUILL (G. ROUTLEDGE. 1845)

"The Philosophers, Politicians, Necromancers, and the learned in every faculty are desired to observe that on January 1, being New Year's Day (oh that we may all lead new lives!), Mr. Newbery intends to publish the following important volumes, bound and gilt, and hereby invites all his little friends who are good to call for them at the Bible and Sun in St. Paul's Churchyard, but those who are naughty to have none." The paper read by Mr. Welsh scarcely fulfils the whole promise of its title, for in place of giving anecdotes of Newbery he refers his listeners to his own volume, "A Bookseller of the Last Century," for fuller details; but what he said in praise of the excellent printing and binding of Newbery's books is well merited. They are, nearly all, comely productions, some with really artistic illustrations, and all marked with care and intelligence which had not hitherto been bestowed on publications intended for juveniles. It is true that most are distinguished for "calculating morality" as the *Athenæum* called it, in re-estimating their merits nearly a century later. It was a period when the advantages of dull moralising were over-prized, when people professed to believe that you could admonish children to a state of perfection which, in their didactic addresses to the small folk, they professed to obey themselves. It was, not to put too fine a point on it, an age of solemn hypocrisy, not perhaps so insincere in intention as in phrase; but, all the same, it repels the more tolerant mood of to-day. Whether or not it be wise to confess to the same frailties and let children know the weaknesses of their elders, it is certainly more honest; and the danger is now rather lest the undue humility of experience should lead children to believe that they are better than their fathers. Probably the honest sympathy now shown to childish ideals is not likely to be misinterpreted, for children are often shrewd judges, and can detect the false from the true, in morals if not in art.

ILLUSTRATION FROM "ROBINSON CRUSOE." BY CHARLES KEENE (JAMES BURNS. 1847)

By 1800 literature for children had become an established fact. Large numbers of publications were ostentatiously addressed to their amusement; but nearly all hid a bitter if wholesome powder in a very small portion of jam. Books of educational purport, like "A Father's Legacy to his Daughter," with reprints of classics that are heavily weighted with morals—Dr. Johnson's "Rasselas" and "Æsop's Fables," for instance—are in the majority. "Robinson Crusoe" is indeed among them, and Bunyan's "Pilgrim's Progress," both, be it noted, books annexed by the young, not designed for them.

The titles of a few odd books which possess more than usually interesting features may be jotted down. Of these, "Little Thumb and the Ogre" (R.

Dutton, 1788), with illustrations by William Blake, is easily first in interest, if not in other respects. Others include "The Cries of London" (1775), "Sindbad the Sailor" (Newbery, 1798), "Valentine and Orson" (Mary Rhynd, Clerkenwell, 1804), "Fun at the Fair" (with spirited cuts printed in red), and Watts's "Divine and Moral Songs," and "An Abridged New Testament," with still more effective designs also in red (Lumsden, Glasgow), "Gulliver's Travels" (greatly abridged, 1815), "Mother Gum" (1805), "Anecdotes of a Little Family" (1795), "Mirth without Mischief," "King Pippin," "The Daisy" (cautionary stories in verse), and the "Cowslip," its companion (with delightfully prim little rhymes that have been reprinted lately). The thirty illustrations in each are by Samuel Williams, an artist who yet awaits his due appreciation. A large number of classics of their kind, "The Adventures of Philip Quarll," "Gulliver's Travels," Blake's "Songs of Innocence," Charles Lamb's "Stories from Shakespeare," Mrs. Sherwood's "Henry and his Bearer," and a host of other religious stories, cannot even be enumerated. But even were it possible to compile a full list of children's books, it would be of little service, for the popular books are in no danger of being forgotten, and the unpopular, as a rule, have vanished out of existence, and except by pure accident could not be found for love or money.

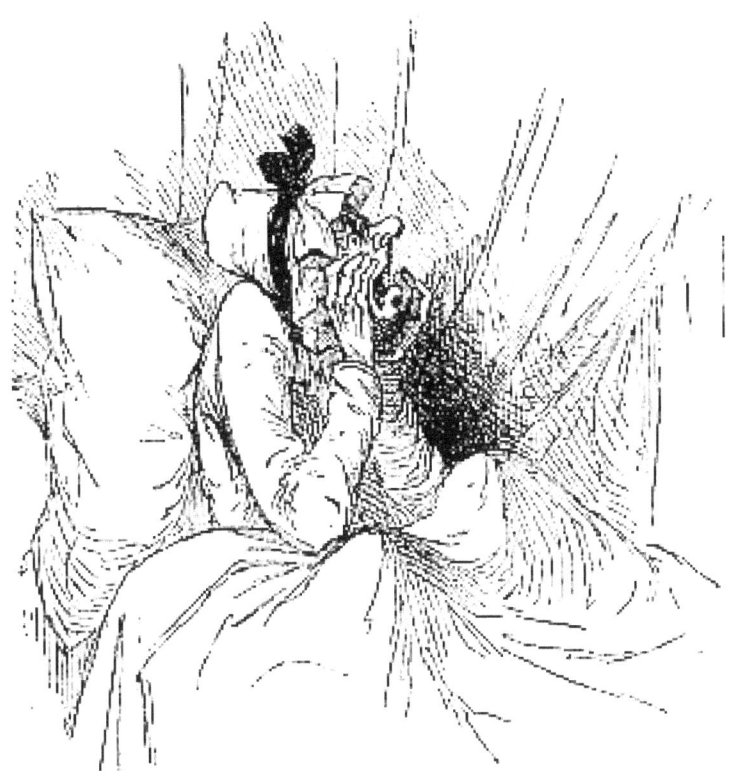

ILLUSTRATION FROM "COMIC NURSERY TALES" (G. ROUTLEDGE. 1846)

With the publications of Newbery and Harris, early in the nineteenth century, we encounter examples more nearly typical of the child's book as we regard it to-day. Among them Harris's "Cabinet" is noticeable. The first four volumes, "The Butterfly's Ball," "The Peacock at Home," "The Lion's Masquerade," and "The Elephant's Ball," were reprinted a few years ago, with the original illustrations by Mulready carefully reproduced. A coloured series of sixty-two books, priced at one shilling and sixpence each (Harris), was extremely popular.

With the "Paths of Learning strewed with Flowers, or English Grammar Illustrated" (1820), we encounter a work not without elegance. Its designs, as we see by the examples reproduced on page 9, are the obvious prototype of Miss Greenaway, the model that inspired her to those dainty trifles which conquered even so stern a critic of modern illustration as Mr. Ruskin. On its cover—a forbidding wrapper devoid of ornament—and repeated within a wreath of roses inside, this preamble occurs: "The purpose of this little book is to obviate the reluctance children evince to the irksome and insipid task of learning the names and meanings of the component parts of grammar. Our intention is to entwine roses with instruction, and however humble our endeavour may appear, let it be recollected that the efforts of a Mouse set the Lion free from his toils." This oddly phrased explanation is typical of the affected geniality of the governess. Indeed, it might have been penned by an assistant to Miss Pinkerton, "the Semiramis of Hammersmith"; if not by that friend of Dr. Johnson, the correspondent of Mrs. Chapone herself, in a moment of gracious effort to bring her intellect down to the level of her pupils.

To us, this hollow gaiety sounds almost cruel. In those days children were always regarded as if, to quote Mark Twain, "every one being born with an equal amount of original sin, the pressure on the square inch must needs be greater in a baby." Poor little original sinners, how very scurvily the world of books and picture-makers treated you less than a century ago! Life for you then was a perpetual reformatory, a place beset with penalties, and echoing with reproofs. Even the literature planned to amuse your leisure was stuck full of maxims and morals; the most piquant story was but a prelude to an awful warning; pictures of animals, places, and rivers failed to conceal undisguised lessons. The one impression that is left by a study of these books is the lack of confidence in their own dignity which papas and mammas betrayed in the early Victorian era. This seems past all doubt when you realise that the common effort of all these pictures and prose is to glorify the impeccable parent, and teach his or her offspring to grovel silently before the stern law-givers who ruled the home.

TITLE-PAGE FROM "THE SCOURING OF THE WHITE HORSE." BY RICHARD DOYLE (MACMILLAN AND CO. 1858)

Of course it was not really so, literature had but lately come to a great middle class who had not learned to be easy; and as worthy folk who talked colloquially wrote in stilted parody of Dr. Johnson's stately periods, so the uncouth address in print to the populace of the nursery was doubtless forgotten in daily intercourse. But the conventions were preserved, and honest fun or full-bodied romance that loves to depict gnomes and hobgoblins, giants and dwarfs in a world of adventure and mystery, was unpopular. Children's books were illustrated entirely by the wonders of the creation, or the still greater wonders of so-called polite society. Never in them, except introduced purposely as an "awful example," do you meet an untidy, careless, normal child. Even the beggars are prim, and the beasts and birds distinctly genteel in their habits. Fairyland was shut to the little ones, who were turned out of their own domain. It seems quite likely that this continued until the German *märchen* (the literary products of Germany were much in favour at this period) reopened the wonderland of the other world about the time that Charles Dickens helped to throw the door still wider. Discovering that the child possessed the right to be amused, the imagination of poets and artists addressed itself at last to the most appreciative of all audiences, a world of newcomers, with insatiable appetites for wonders real and imaginary.

ILLUSTRATION (REDUCED) FROM "MISUNDERSTOOD" BY GEORGE DU MAURIER (RICHARD BENTLEY AND SON. 1874)

But for many years before the Victorian period folklore was left to the peasants, or at least kept out of reach of children of the higher classes. No doubt old nurses prattled it to their charges, perhaps weak-minded mothers occasionally repeated the ancient legends, but the printing-press set its face against fancy, and offered facts in its stead. In the list of sixty-two books before mentioned, if we except a few nursery jingles such as "Mother Hubbard" and "Cock Robin," we find but two real fairy stories, "Cinderella," "Puss-in-Boots," and three old-world narratives of adventure, "Whittington and His Cat," "The Seven Champions of Christendom," and "Valentine and Orson." The rest are "Peter Piper's Practical Principles of Plain and Perfect Pronunciation," "The Monthly Monitor," "Tommy Trip's Museum of Beasts," "The Perambulations of a Mouse," and so on, with a few things like "The House that Jack Built," and "A, Apple Pie," that are but daily facts put into story shape. Now it is clear that the artists inspired by fifty of these had no chance of displaying their imagination, and every opportunity of pointing a moral; and it is painful to be obliged to own that they succeeded beyond belief in their efforts to be dull. Of like sort are "A Visit to the Bazaar" (Harris, 1814), and "The Dandies' Ball" (1820).

ILLUSTRATION FROM "THE PRINCESS AND THE GOBLIN." (STRAHAN. 1871. NOW PUBLISHED BY BLACKIE AND SON)

Nor must we forget a work very popular at this period, "Keeper in Search of His Master," although its illustrations are not its chief point.

According to a very interesting preface Mr. Andrew Tuer contributed to "The Leadenhall Series of Reprints of Forgotten Books for Children in 1813," "Dame Wiggins of Lee" was first issued by A. K. Newman and Co. of the Minerva Press. This book is perhaps better known than any of its date owing to Mr. Ruskin's reprint with additional verses by himself, and new designs by Miss Kate Greenaway supplementing the original cuts, which were re-engraved in facsimile by Mr. Hooper. Mr. Tuer attributes the design of these latter to R. Stennet (or Sinnet?), who illustrated also "Deborah Dent and her Donkey" and "Madame Figs' Gala." Newman issued many of these books, in conjunction with Messrs. Dean and Mundy, the direct ancestors of the firm of Dean and Son, still flourishing, and still engaged in providing cheap and attractive books for children. "The Gaping Wide-mouthed Waddling Frog" is another book of about this period, which Mr. Tuer included in his reprints. Among the many illustrated volumes which bear the imprint of A. K. Newman, and Dean and Mundy, are "A, Apple Pie," "Aldiborontiphoskyphorniostikos," "The House that Jack Built," "The Parent's Offering for a Good Child" (a very pompous and irritating series of dialogues), and others that are even more directly educational. In all these the engravings are in fairly correct outline, coloured with four to six washes of showy crimson lake, ultramarine, pale green, pale sepia, and gamboge.

ILLUSTRATION FROM "GUTTA PERCHA WILLIE." BY ARTHUR HUGHES (STRAHAN. 1870. NOW PUBLISHED BY BLACKIE AND SON)

ILLUSTRATION FROM "AT THE BACK OF THE NORTH WIND." BY ARTHUR HUGHES (STRAHAN. 1869. NOW PUBLISHED BY BLACKIE AND SON)

Even the dreary text need not have made the illustrators quite so dull, as we know that Randolph Caldecott would have made an illustrated "Bradshaw" amusing; but most of his earlier predecessors show no less power in making anything they touched "un-funny." Nor as art do their pictures interest you any more than as anecdotes.

Of course the cost of coloured engravings prohibited their lavish use. All were tinted by hand, sometimes with the help of stencil plates, but more often by brush. The print colourers, we are told, lived chiefly in the Pentonville district, or in some of the poorer streets near Leicester Square. A few survivors are still to be found; but the introduction first of lithography, and later of photographic processes, has killed the industry, and even the most fanatical apostle of the old crafts cannot wish the "hand-painter" back again. The outlines were either cut on wood, as in the early days of printing until the present, or else engraved on metal. In each case all colour was painted afterwards, and in scarce a single instance (not even in the

Rowlandson caricatures or patriotic pieces) is there any attempt to obtain an harmonious scheme such as is often found in the tinted mezzo-tints of the same period.

ILLUSTRATION FROM "AT THE BACK OF THE NORTH WIND." BY ARTHUR HUGHES (STRAHAN. 1869. NOW PUBLISHED BY BLACKIE AND SON)

Of works primarily intended for little people, an "Hieroglyphical Bible" for the amusement and instruction of the younger generation (1814) may be noted. This was a mixture of picture-puns and broken words, after the fashion of the dreary puzzles still published in snippet weeklies. It is a melancholy attempt to turn Bible texts to picture puzzles, a book permitted by the unco' guid to children on wet Sunday afternoons, as some younger members of large families, whose elder brothers' books yet lingered forty or even fifty years after publication, are able to endorse with vivid and depressed remembrance. Foxe's "Book of Martyrs" and Bunyan's "Pilgrim's Progress" are of the same type, and calculated to fill a nervous child with grim terrors, not lightened by Watts's "Divine and Moral Songs," that gloated on the dreadful hell to which sinful children were doomed, "with devils in darkness, fire and chains." But this painful side of the subject is not to be discussed here. Luckily the artists—except in the "grown-up" books referred to—disdained to enforce the terrors of Dr. Watts, and pictured less horrible themes.

With Cruikshank we encounter almost the first glimpse of the modern ideal. His "Grimm's Fairy Tales" are delightful in themselves, and marvellous in comparison with all before, and no little after.

ILLUSTRATION FROM "THE LITTLE WONDER HORN." BY J. MAHONEY (H. S. KING AND CO. 1872. GRIFFITH AND FARRAN. 1887)

These famous illustrations to the first selection of Grimm's "German Popular Stories" appeared in 1824, followed by a second series in 1826. Coming across this work after many days spent in hunting up children's books of the period, the designs flashed upon one as masterpieces, and for the first time seemed to justify the great popularity of Cruikshank. For their vigour and brilliant invention, their *diablerie* and true local colour, are amazing when contrasted with what had been previously. Wearied of the excessive eulogy bestowed upon Cruikshank's illustrations to Dickens, and unable to accept the artist as an illustrator of real characters in fiction, when he studies his elfish and other-worldly personages, the most grudging critic must needs yield a full tribute of praise. The volumes (published by Charles Tilt, of 82 Fleet Street) are extremely rare; for many years past the sale-room has recorded fancy prices for all Cruikshank's illustrations, so that a lover of modern art has been jealous to note the amount paid for by many extremely poor pictures by this artist, when even original drawings for the masterpieces by later illustrators went for a song. In Mr. Temple Scott's indispensable "Book Sales of 1896" we find the two volumes (1823-6) fetched £12 12*s*.

"IN NOOKS WITH BOOKS"
AN AUTO-LITHOGRAPH BY
R. ANNING BELL

These must not be confounded with Cruikshank's "Fairy Library" (1847-64), a series of small books in paper wrappers, now exceedingly rare, which are more distinctly prepared for juvenile readers. The illustrations to these do not rise above the level of their day, as did the earlier ones. But this is owing largely to the fact that the standard had risen far above its old average in the thirty years that had elapsed. Amid the mass of volumes illustrated by Cruikshank comparatively few are for juveniles; some of these are: "Grimm's Gammer Grethel"; "Peter Schlemihl" (1824); "Christmas Recreation" (1825); "Hans of Iceland" (1825); "German Popular Stories" (1823); "Robinson Crusoe" (1831); "The Brownies" (1870); "Loblie-by-the-Fire" (1874); "Tom Thumb" (1830); and "John Gilpin" (1828).

ILLUSTRATION FROM "SPEAKING LIKENESSES." BY
ARTHUR HUGHES
(MACMILLAN AND CO. 1874)

The works of Richard Doyle (1824-1883) enjoy in a lesser degree the sort of inflated popularity which has gathered around those of Cruikshank. With much spirit and pleasant invention, Doyle lacked academic skill, and often betrays considerable weakness, not merely in composition, but in invention. Yet the qualities which won him reputation are by no means despicable. He evidently felt the charm of fairyland, and peopled it with droll little folk who are neither too human nor too unreal to be attractive. He joined the staff of *Punch* when but nineteen, and soon, by his political cartoons, and his famous "Manners and Customs of ye English drawn from ye Quick," became an established favourite. His design for the cover of *Punch* is one of his happiest inventions. So highly has he been esteemed that the National Gallery possesses one of his pictures, *The Triumphant Entry; a Fairy Pageant*. Children's books with his illustrations are numerous; perhaps the most important are "The Enchanted Crow" (1871), "Feast of Dwarfs" (1871), "Fortune's Favourite" (1871), "The Fairy Ring" (1845), "In Fairyland" (1870), "Merry Pictures" (1857), "Princess Nobody" (1884), "Mark Lemon's Fairy Tales" (1868), "A Juvenile Calendar" (1855), "Fairy Tales from all Nations" (1849), "Snow White and Rosy Red" (1871), Ruskin's "The King of the Golden River" (1884), Hughes's "Scouring of the White Horse" (1859), "Jack the Giant Killer" (1888), "Home for the Holidays" (1887), "The Whyte Fairy Book" (1893). The three last are, of course, posthumous publications.

Still confining ourselves to the pre-Victorian period, although the works in question were popular several decades later, we find "Sandford and Merton" (first published in 1783, and constantly reprinted), "The Swiss Family Robinson," the beginning of "Peter Parley's Annals," and a vast number of other books with the same pseudonym appended, and a host of didactic works, a large number of which contained pictures of animals and other natural objects, more or less well drawn. But the pictures in these are not of any great consequence, merely reflecting the average taste of the day, and very seldom designed from a child's point of view.

ILLUSTRATION FROM "UNDINE." BY SIR JOHN TENNIEL
(JAMES BURNS. 1845)

This very inadequate sketch of the books before 1837 is not curtailed for want of material, but because, despite the enormous amount, very few show attempts to please the child; to warn, to exhort, or to educate are their chief aims. Occasionally a Bewick or an artist of real power is met with, but the bulk is not only dull, but of small artistic value. That the artist's name is rarely given must not be taken as a sign that only inept draughtsmen were employed, for in works of real importance up to and even beyond this date we often find his share ignored. After a time the engraver claims to be considered, and by degrees the designer is also recognised; yet for the most part illustration was looked upon merely as "jam" to conceal the pill. The old Puritan conception of art as vanity had something to do with this, no doubt; for adults often demand that their children shall obey a sterner rule of life than that which they accept themselves.

ILLUSTRATION FROM "ELLIOTT'S NURSERY RHYMES" BY W. J. WIEGAND
(NOVELLO, 1870)

Before passing on, it is as well to summarise this preamble and to discover how far children's books had improved when her Majesty came to the throne. The old woodcut, rough and ill-drawn, had been succeeded by the masterpieces of Bewick, and the respectable if dull achievements of his followers. In the better class of books were excellent designs by artists of some repute fairly well engraved. Colouring by hand, in a primitive fashion, was applied to these prints and to impressions from copperplates. A certain prettiness was the highest aim of most of the latter, and very few were designed only to amuse a child. It seems as if all concerned were bent on unbending themselves, careful to offer grains of truth to young minds with an occasional terrible falsity of their attitude; indeed, its satire and profound analysis make it superfluous to reopen the subject. As one might expect, the literature, "genteel" and dull, naturally desired pictures in the same key. The art of even the better class of children's books was satisfied if it succeeded in being "genteel," or, as Miss Limpenny would say, "cumeelfo." Its ideal reached no higher, and sometimes stopped very far below that modest standard. This is the best (with the few exceptions already noted) one can say of pre-Victorian illustration for children.

ILLUSTRATION FROM "ELLIOTT'S NURSERY RHYMES" BY H. STACY MARKS, R.A. (NOVELLO. 1870)

If there is one opinion deeply rooted in the minds of the comparatively few Britons who care for art, it is a distrust of "The Cole Gang of South Kensington;" and yet if there be one fact which confronts any student of the present revival of the applied arts, it is that sooner or later you come to its first experiments inspired or actually undertaken by Sir Henry Cole. Under the pseudonym of "Felix Summerley" we find that the originator of a hundred revivals of the applied arts, projected and issued a series of children's books which even to-day are decidedly worth praise. It is the fashion to trace everything to Mr. William Morris, but in illustrations for children as in a hundred others "Felix Summerley" was setting the ball rolling when Morris and the members of the famous firm were schoolboys.

ILLUSTRATION FROM "THE WATER BABIES" BY SIR R. NOEL PATON (MACMILLAN AND CO. 1863)

To quote from his own words: "During this period (*i.e.*, about 1844), my young children becoming numerous, their wants induced me to publish a rather long series of books, which constituted 'Summerley's Home Treasury,' and I had the great pleasure of obtaining the welcome assistance of some of the first artists of the time in illustrating them—Mulready, R.A., Cope, R.A., Horsley, R.A., Redgrave, R.A., Webster, R.A., Linnell and his three sons, John, James, and William, H. J. Townsend, and others.... The preparation of these books gave me practical knowledge in the technicalities of the arts of type-printing, lithography, copper and steel-plate engraving and printing, and bookbinding in all its varieties in metal, wood, leather, &c."

Copies of the books in question appear to be very rare. It is doubtful if the omnivorous British Museum has swallowed a complete set; certainly at the Art Library of South Kensington Museum, where, if anywhere, we might expect to find Sir Henry Cole completely represented, many gaps occur.

ILLUSTRATION FROM "THE ROYAL UMBRELLA." BY LINLEY SAMBOURNE (GRIFFITH AND FARRAN. 1880)

How far Mr. Joseph Cundall, the publisher, should be awarded a share of the credit for the enterprise is not apparent, but his publications and writings, together with the books issued later by Cundall and Addey, are all marked with the new spirit, which so far as one can discover was working in many minds at this time, and manifested itself most conspicuously through the Pre-Raphaelites and their allies. This all took place, it must be remembered, long before 1851. We forget often that if that exhibition has any important place in the art history of Great Britain, it does but prove that much preliminary work had been already accomplished. You cannot exhibit what does not exist; you cannot even call into being "exhibition specimens" at a few months notice, if something of the same sort, worked for ordinary commerce, has not already been in progress for years previously.

ILLUSTRATION FROM "ON A PINCUSHION." BY WILLIAM DE MORGAN (SEELEY, JACKSON AND HALLIDAY. 1877)

Almost every book referred to has been examined anew for the purposes of this article. As a whole they might fail to impress a critic not peculiarly interested in the matter. But if he tries to project himself to the period that produced them, and realises fully the enormous importance of first efforts, he will not estimate grudgingly their intrinsic value, but be inclined to credit them with the good things they never dreamed of, as well as those they tried to realise and often failed to achieve. Here, without any prejudice for or against the South Kensington movement, it is but common justice to record Sir Henry Cole's share in the improvement of children's books; and later on his efforts on behalf of process engraving must also not be forgotten.

To return to the books in question, some extracts from the original prospectus, which speaks of them as "purposed to cultivate the Affections, Fancy, Imagination, and Taste of Children," are worth quotation:

"The character of most children's books published during the last quarter of a century, is fairly typified in the name of Peter Parley, which the writers of some hundreds of them have assumed. The books themselves have been addressed after a narrow fashion, almost entirely to the cultivation of the understanding of children. The many tales sung or said from time to time immemorial, which appealed to the other, and certainly not less important elements of a little child's mind, its fancy, imagination, sympathies, affections, are almost all gone out of memory, and are scarcely to be obtained. 'Little Red Riding Hood,' and other fairy tales hallowed to children's use, are now turned into ribaldry as satires for men; as for the creation of a new fairy tale or touching ballad, such a thing is unheard of. That the influence of all this is hurtful to children, the conductor of this series firmly believes. He has practical experience of it every day in his own family, and he doubts not that there are many others who entertain the same opinions as himself. He purposes at least to give some evidence of his belief, and to produce a series of works, the character of which may be briefly described as anti-Peter Parleyism.

ILLUSTRATION FROM "THE NECKLACE OF PRINCESS
FIORIMONDE."
BY WALTER CRANE (MACMILLAN AND CO. 1880)

"Some will be new works, some new combinations of old materials, and some reprints carefully cleared of impurities, without deterioration to the points of the story. All will be illustrated, but not after the usual fashion of children's books, in which it seems to be assumed that the lowest kind of art is good enough to give first impressions to a child. In the present series, though the statement may perhaps excite a smile, the illustrations will be selected from the works of Raffaelle, Titian, Hans Holbein, and other old masters. Some of the best modern artists have kindly promised their aid in creating a taste for beauty in little children." Did space permit, a selection from the reviews of the chief literary papers that welcomed the new venture would be instructive. There we should find that even the most cautious critic, always "hedging" and playing for safety, felt compelled to accord a certain amount of praise to the new enterprise.

It is true that "Felix Summerley" created only one type of the modern book. Possibly the "stories turned into satires" to which he alludes are the entirely amusing volumes by F. H. Bayley, the author of "A New Tale of a Tub." As it happened that these volumes were my delight as a small boy, possibly I am unduly fond of them; but it seems to me that their humour—*à la* Ingoldsby, it is true—and their exuberantly comic drawings, reveal the first glimpses of lighter literature addressed specially to children, that long after found its masterpieces in the "Crane" and "Greenaway" and "Caldecott" Toy Books, in "Alice in Wonderland," and in a dozen other treasured volumes, which are now classics. The chief claim for the Home Treasury series to be considered as the advance guard of our present sumptuous volumes, rests not so much upon the quality of their designs or the brightness of their literature. Their chief importance is that in each of them we find for the first time that the externals of a child's book are most carefully considered. Its type is well chosen, the proportions of its page are evidently studied, its binding, even its end-papers, show that some one person was doing his best to attain perfection. It is this conscious effort, whatever it actually realised, which distinguishes the result from all before.

It is evident that the series—the Home Treasury—took itself seriously. Its purpose was Art with a capital A—a discovery, be it noted, of this period. Sir Henry Cole, in a footnote to the very page whence the quotation above was extracted, discusses the first use of "Art" as an adjective denoting the *Fine* Arts.

ILLUSTRATION FROM "HOUSEHOLD STORIES FROM GRIMM." BY WALTER CRANE (MACMILLAN AND CO. 1882)

Here it is more than ever difficult to keep to the thread of this discourse. All that South Kensington did and failed to do, the æsthetic movement of the eighties, the new gospel of artistic salvation by Liberty fabrics and De Morgan tiles, the erratic changes of fashion in taste, the collapse of Gothic architecture, the triumph of Queen Anne, and the Arts and Crafts movement of the nineties—in short, all the story of Art in the last fifty years, from the new Law Courts to the Tate Gallery, from Felix Summerley to a Hollyer photograph, from the introduction of glyptography to the pictures in the *Daily Chronicle*, demand notice. But the door must be shut on the turbulent throng, and only children's books allowed to pass through.

The publications by "Felix Summerley," according to the list in "Fifty Years of Public Work," by Sir Henry Cole, K.C.B. (Bell, 1884), include: "Holbein's Bible Events," eight pictures, coloured by Mr. Linnell's sons, 4*s.* 6*d.*; "Raffaelle's Bible Events," six pictures from the Loggia, drawn on stone by Mr. Linnell's children and coloured by them, 5*s.* 6*d.*; "Albert Dürer's Bible Events," six pictures from Dürer's "Small Passion," coloured by the brothers Linnell; "Traditional Nursery Songs," containing eight pictures; "The Beggars coming to Town," by C. W. Cope, R.A.; "By, O my Baby!" by R. Redgrave, R.A.; "Mother Hubbard," by T. Webster, R.A.; "1, 2, 3, 4, 5," "Sleepy Head," "Up in a Basket," "Cat asleep by the Fire," by John Linnell, 4*s.* 6*d.*, coloured; "The Ballad of Sir Hornbook," by Thos. Love Peacock, with eight pictures by H. Corbould, coloured, 4*s.* 6*d.* (A book with the same title, also described as a "grammatico-allegorical ballad," was published by N. Haites in 1818.) "Chevy Chase," with music and four pictures by Frederick Tayler, President of the Water-Colour Society, coloured, 4*s.* 6*d.*; "Puck's Reports to Oberon"; Four new Faëry Tales: "The Sisters," "Golden Locks," "Grumble and Cherry," "Arts and Arms," by C. A. Cole, with six pictures by J. H. Townsend, R. Redgrave, R.A., J. C. Horsley, R.A., C. W. Cope, R.A., and F. Tayler; "Little Red Riding Hood," with four pictures by Thos. Webster, coloured, 3*s.* 6*d.*; "Beauty and the Beast," with four pictures by J. C. Horsley, R.A., coloured, 3*s.* 6*d.*; "Jack and the Bean Stalk," with four pictures by C. W. Cope, R.A., coloured, 3*s.* 6*d.*; "Cinderella," with four pictures by E. H. Wehnert, coloured, 3*s.* 6*d.*; "Jack the Giant Killer," with four pictures by C. W. Cope, coloured, 3*s.* 6*d.*; "The Home Treasury Primer," printed in colours, with drawing on zinc, by W. Mulready, R.A.; "Alphabets of Quadrupeds," selected from the works of Paul Potter, Karl du Jardin, Teniers, Stoop, Rembrandt, &c., and drawn from nature; "The Pleasant History of Reynard the Fox," with forty of the fifty-seven etchings made by Everdingen in 1752, coloured, 31*s.* 6*d.*; "A Century of Fables," with pictures by the old masters.

ILLUSTRATION FROM "A WONDER BOOK FOR GIRLS AND BOYS." BY WALTER CRANE (OSGOOD, MCILVAINE AND CO. 1892)

To this list should be added—if it is not by "Felix Summerley," it is evidently conceived by the same spirit and published also by Cundall—"Gammer Gurton's Garland," by Ambrose Merton, with illustrations by T. Webster and others. This was also issued as a series of sixpenny books, of which Mr. Elkin Mathews owns a nearly complete set, in their original covers of gold and coloured paper.

It would be very easy to over-estimate the intrinsic merit of these books, but when you consider them as pioneers it would be hard to over-rate the importance of the new departure. To enlist the talent of the most popular artists of the period, and produce volumes printed in the best style of the Chiswick Press, with bindings and end-papers specially designed, and the whole "get up" of the book carefully considered, was certainly a bold innovation in the early forties. That it failed to be a profitable venture one may deduce from the fact that the "Felix Summerley" series did not run to many volumes, and that the firm who published them, after several changes, seems to have expired, or more possibly was incorporated with some other venture. The books themselves are forgotten by most booksellers to-day, as I have discovered from many fruitless demands for copies.

The little square pamphlets by F. H. Bayley, to which allusion has already been made, include "Blue Beard;" "Robinson Crusoe," and "Red Riding Hood," all published about 1845-6.

ILLUSTRATION FROM "THE QUEEN OF THE PIRATE ISLE."
BY KATE GREENAWAY (EDMUND EVANS. 1887)

Whether "The Sleeping Beauty," then announced as in preparation, was published, I do not know. Their rhyming chronicle in the style of the "Ingoldsby Legends" is neatly turned, and the topical allusions, although out of date now, are not sufficiently frequent to make it unintelligible. The pictures (possibly by Alfred Crowquill) are conceived in a spirit of burlesque, and are full of ingenious conceits and no little grim vigour. The design of Robinson Crusoe roosting in a tree—

And so he climbs up a very tall tree,
And fixes himself to his comfort and glee,
Hung up from the end of a branch by his breech,
Quite out of all mischievous quadrupeds' reach.
A position not perfectly easy 't is true,
But yet at the same time consoling and new—

reproduced on p. 13, shows the wilder humour of the illustrations. Another of Blue Beard, and one of the wolf suffering from undigested grandmother, are also given. They need no comment, except to note that in the originals, printed on a coloured tint with the high lights left white, the ferocity of Blue Beard is greatly heightened. The wolf, "as he lay there brimful of grandmother and guilt," is one of the best of the smaller pictures in the text.

Other noteworthy books which appeared about this date are Mrs. Felix Summerley's "Mother's Primer," illustrated by W. M[ulready?], Longmans, 1843; "Little Princess," by Mrs. John Slater, 1843, with six charming lithographs by J. C. Horsley, R.A. (one of which is reproduced on p. 11); the "Honey Stew," of the Countess Bertha Jeremiah How, 1846, with coloured plates by Harrison Weir; "Early Days of English Princes," with capital illustrations by John Franklin; and a series of Pleasant Books for Young Children, 6*d.* plain and 1*s.* coloured, published by Cundall and Addey.

ILLUSTRATION FROM "LITTLE FOLKS" BY KATE GREENAWAY
(CASSELL AND CO.)

In 1846 appeared a translation of De La Motte Fouqué's romances, "Undine" being illustrated by John Tenniel, jun., and the following volumes by J. Franklin, H. C. Selous, and other artists. The Tenniel designs, as the frontispiece reproduced on p. 20 shows clearly, are interesting both in themselves and as the earliest published work of the famous *Punch* cartoonist. The strong German influence they show is also apparent in nearly all the decorations. "The Juvenile Verse and Picture Book" (1848), also contains designs by Tenniel, and others by W. B. Scott and Sir John Gilbert. The ideal they established is maintained more or less closely for a long period. "Songs for Children" (W. S. Orr, 1850); "Young England's Little Library" (1851); Mrs. S. C. Hall's "Number One," with pictures by John Absolon (1854); "Stories about Dogs," with "plates by Thomas Landseer" (Bogue, *c.* 1850); "The Three Bears," illustrated by Absolon and Harrison Weir (Addey and Co., no date); "Nursery Poetry" (Bell and Daldy, 1859), may be noted as typical examples of this period.

ILLUSTRATION FROM "THE PIED PIPER OF HAMELIN" BY
KATE GREENAWAY (EDMUND EVANS)

In "Granny's Story Box" (Piper, Stephenson, and Spence, about 1855), a most delicious collection of fairy tales illustrated by J. Knight, we find the author in his preface protesting against the opinion of a supposititious old lady who "thought all fairy tales were abolished years ago by Peter Parley and the *Penny Magazine.*" These fanciful stories deserve to be republished, for they are not old-fashioned, even if their pictures are.

To what date certain delightfully printed little volumes, issued by Tabart and Co., 157 Bond Street, may be ascribed I know not—probably some years before the time we are considering, but they must not be overlooked. The title of one, "Mince Pies for Christmas," suggests that it is not very far before, for the legend of Christmas festivities had not long been revived for popular use.

"The Little Lychetts," by the author of "John Halifax," illustrated by Henry Warren, President of the New Society of Painters in Water-Colours (now the R.I.) is remarkable for the extremely uncomely type of children it depicts; yet that its charm is still vivid, despite its "severe" illustrations, you have but to lend it to a child to be convinced quickly.

"Jack's Holiday," by Albert Smith (undated), suggests a new field of research which might lead us astray, as Smith's humour is more often addressed primarily to adults. Indeed, the effort to make this chronicle even representative, much less exhaustive, breaks down in the fifties, when so much good yet not very exhilarating material is to be found in every publisher's list. John Leech in "The Silver Swan" of Mdme. de Chatelaine; Charles Keene in "The Adventures of Dick Bolero" (Darton, no date), and "Robinson Crusoe" (drawn upon for illustration here), and others of the *Punch* artists, should find their works duly catalogued even in this hasty sketch; but space compels scant justice to many artists of the period, yet if the most popular are left unnoticed such omission will more easily right itself to any reader interested in the subject.

Many show influences of the Gothic revival which was then in the air, but only those which have some idea of book decoration as opposed to inserted pictures. For a certain "formal" ornamentation of the page was in fashion in the "forties" and "fifties," even as it is to-day.

ILLUSTRATION FROM "CAPE TOWN DICKY" BY ALICE
HAVERS
(C. W. FAULKNER AND CO.)

To the artists named as representative of this period one must not forget to add Mr. Birket Foster, who devoted many of his felicitous studies of English pastoral life to the adornment of children's books. But speaking broadly of the period from the Queen's Accession to 1865, except that the subjects are of a sort supposed to appeal to young minds, their conception differs in no way from the work of the same artists in ordinary literature. The vignettes of scenery have childish instead of grown-up figures in the foregrounds; the historical or legendary figures are as seriously depicted in the one class of books as in the other. Humour is conspicuous by its absence—or, to be more accurate, the humour is more often in the accompanying anecdote than in the picture. Probably if the authorship of hundreds of the illustrations of "Peter Parley's Annuals" and other books of this period could be traced, artists as famous as Charles Keene might be found to have contributed. But,

- 58 -

owing to the mediocre wood-engraving employed, or to the poor printing, the pictures are singularly unattractive. As a rule, they are unsigned and appear to be often mere pot-boilers—some no doubt intentionally disowned by the designer—others the work of 'prentice hands who afterwards became famous. Above all they are, essentially, illustrations to children's books only because they chanced to be printed therein, and have sometimes done duty in "grown-up" books first. Hence, whatever their artistic merits, they do not appeal to a student of our present subject. They are accidentally present in books for children, but essentially they belong to ordinary illustrations.

Indeed, speaking generally, the time between "Felix Summerley" and *Walter Crane*, which saw two Great Exhibitions and witnessed many advances in popular illustration, was too much occupied with catering for adults to be specially interested in juveniles. Hence, notwithstanding the names of "illustrious illustrators" to be found on their title-pages, no great injustice will be done if we leave this period and pass on to that which succeeded it. For the Great Exhibition fostered the idea that a smattering of knowledge of a thousand and one subjects was good. Hence the chastened gaiety of its mildly technical science, its popular manuals by Dr. Dionysius Lardner, and its return in another form to the earlier ideal that amusement should be combined with instruction. All sorts of attempts were initiated to make Astronomy palatable to babies, Botany an amusing game for children, Conchology a parlour pastime, and so on through the alphabet of sciences down to Zoology, which is never out of favour with little ones, even if its pictures be accompanied by a dull encylopædia of fact.

ILLUSTRATION FROM "THE WHITE SWANS" BY ALICE HAVERS
(By permission of Mr. Albert Hildesheimer)

Therefore, except so far as the work of certain illustrators, hereafter noticed, touches this period, we may leave it; not because it is unworthy of most serious attention, for in Sir John Gilbert, Birket Foster, Harrison Weir, and the rest, we have men to reckon with whenever a chronicle of English illustration is in question, but only because they did not often feel disposed to make their work merely amusing. In saying this it is not suggested that they should have tried to be always humorous or archaic, still less to bring down their talent to the supposed level of a child; but only to record the fact that they did not. For instance, Sir John Gilbert's spirited compositions to a "Boy's Book of Ballads" (Bell and Daldy) as you see them mixed with other of the master's work in the reference scrap-books of the publishers, do not at once separate themselves from the rest as "juvenile" pictures.

Nor as we approach the year 1855 (of the "Music Master"), and 1857 (when the famous edition of Tennyson's Poems began a series of superbly illustrated books), do we find any immediate change in the illustration of children's books. The solitary example of Sir Edward Burne-Jones's efforts in this direction, in the frontispiece and title-page to Maclaren's "The Fairy Family" (Longmans, 1857), does not affect this statement. But soon after, as the school of Walker and Pinwell became popular, there is a change in books of all sorts, and Millais and Arthur Hughes, two of the three illustrators of the notable "Music Master," come into our list of children's artists. At this point the attempt to weave a chronicle of children's books somewhat in the date of their publication must give way to a desultory notice of the most prominent illustrators. For we have come to the beginning of to-day rather than the end of yesterday, and can regard the "sixties" onwards as part of the present.

ILLUSTRATION FROM "THE RED FAIRY BOOK." BY
LANCELOT SPEED (LONGMANS, GREEN AND CO.)

It is true that the Millais of the wonderful designs to "The Parables" more often drew pictures of children than of children's pet themes, but all the same they are entirely lovable, and appeal equally to children of all ages. But his work in this field is scanty; nearly all will be found in "Little Songs for me to Sing" (Cassell), or in "Lilliput Levee" (1867), and these latter had appeared previously in *Good Words*. Of Arthur Hughes's work we will speak later.

Another artist whose work bulks large in our subject—Arthur Boyd Houghton—soon appears in sight, and whether he depicted babies at play as in "Home Thoughts and Home Scenes," a book of thirty-five pictures of little people, or imagined the scenes of stories dear to them in "The Arabian Nights," or books like "Ernie Elton" or "The Boy Pilgrims," written especially for them, in each he succeeded in winning their hearts, as every one must admit who chanced in childhood to possess his work. So much has been printed lately of the artist and his work, that here a bare reference will suffice.

ILLUSTRATION FROM "THE RED FAIRY BOOK." BY LANCELOT SPEED (LONGMANS, GREEN AND CO.)

ILLUSTRATION FROM "THE RED FAIRY BOOK." BY
LANCELOT SPEED (LONGMANS, GREEN AND CO.)

Arthur Hughes, whose work belongs to many of the periods touched upon in this rambling chronicle, may be called *the* children's "black-and-white" artist of the "sixties" (taking the date broadly as comprising the earlier "seventies" also), even as Walter Crane is their "limner in colours." His work is evidently conceived with the serious make-believe that is the very essence of a child's imagination. He seems to put down on paper the very spirit of fancy. Whether as an artist he is fully entitled to the rank some of his admirers (of whom I am one) would claim, is a question not worth raising here—the future will settle that for us. But as a children's illustrator he is surely illustrator-in-chief to the Queen of the Fairies, and to a whole generation of readers of "Tom Brown's Schooldays" also. His contributions to "Good Words for the Young" would alone entitle him to high eminence. In addition to these, which include many stories perhaps better known in book form, such as: "The Boy in Grey" (H. Kingsley), George Macdonald's "At the Back of the North Wind," "The Princess and the Goblin," "Ranald Bannerman's Boyhood," "Gutta-Percha Willie" (these four were published by Strahan, and now may be obtained in reprints issued by Messrs. Blackie), and "Lilliput Lectures" (a book of essays for children by Matthew Browne), we find him as sole illustrator of Christina Rossetti's "Sing Song," "Five Days' Entertainment at Wentworth Grange," "Dealings with the Fairies," by George Macdonald (a very scarce volume nowadays), and the chief contributor to the first illustrated edition of "Tom Brown's Schooldays." In Novello's "National Nursery Rhymes" are also several of his designs.

This list, which occupies so small a space, represents several hundred designs, all treated in a manner which is decorative (although it eschews the Dürer line), but marked by strong "colour." Indeed, Mr. Hughes's technique is all his own, and if hard pressed one might own that in certain respects it is not impeccable. But if his textures are not sufficiently differentiated, or even if his drawing appears careless at times—both charges not to be admitted without vigorous protest—granting the opponent's view for the moment, it would be impossible to find the same peculiar tenderness and naïve fancy in the work of any other artist. His invention seems inexhaustible and his composition singularly fertile: he can create "bogeys" as well as "fairies."

ILLUSTRATION FROM "DOWN THE SNOW STAIRS." BY
GORDON BROWNE (BLACKIE AND SON)

It is true that his children are related to the sexless idealised race of Sir Edward Burne-Jones's heroes and heroines; they are purged of earthy taint, and idealised perhaps a shade too far. They adopt attitudes graceful if not realistic, they have always a grave serenity of expression; and yet withal they endear themselves in a way wholly their own. It is strange that a period which has bestowed so much appreciation on the work of the artists of "the sixties" has seen no knight-errant with "Arthur Hughes" inscribed on his banner—no exhibition of his black-and-white work, no craze in auction-rooms for first editions of books he illustrated. He has, however, a steady if limited band of very faithful devotees, and perhaps—so inconsistent are we all—they love his work all the better because the blast of popularity has not trumpeted its merits to all and sundry.

Three artists, often coupled together—Walter Crane, Randolph Caldecott, and Kate Greenaway—have really little in common, except that they all designed books for children which were published about the same period. For Walter Crane is the serious apostle of art for the nursery, who strove to beautify its ideal, to decorate its legends with a real knowledge of architecture and costume, and to "mount" the fairy stories with a certain archæological splendour, as Sir Henry Irving has set himself to mount Shakespearean drama. Caldecott was a fine literary artist, who was able to express himself with rare facility in pictures in place of words, so that his comments upon a simple text reveal endless subtleties of thought. Indeed, he continued to make a fairly logical sequence of incidents out of the famous nonsense paragraph invented to confound mnemonics by its absolute irrelevancy. Miss Greenaway's charm lies in the fact that she first recognised quaintness in what had been considered merely "old fashion," and continued to infuse it with a glamour that made it appear picturesque. Had she dressed her figures in contemporary costume most probably her work would have taken its place with the average, and never obtained more than common popularity.

ILLUSTRATION FROM "ROBINSON CRUSOE" BY GORDON
BROWNE
(BLACKIE AND SON)

But Mr. Walter Crane is almost unique in his profound sympathy with the fantasies he imagines. There is no trace of make-believe in his designs. On the contrary, he makes the old legends become vital, not because of the personalities he bestows on his heroes and fairy princesses—his people move often in a rapt ecstasy—but because the adjuncts of his *mise-en-scènes* are realised intimately. His prince is much more the typical hero than any particular person; his fair ladies might exchange places, and few would notice the difference; but when it comes to the environment, the real incidents of the story, then no one has more fully grasped both the dramatic force and the local colour. If his people are not peculiarly alive, they are in harmony with the re-edified cities and woods that sprang up under his pencil. He does not bestow the hoary touch of antiquity on his mediæval buildings; they are all new and comely, in better taste probably than the actual buildings, but not more idealised than are his people. He is the true artist of fairyland, because he recognises its practical possibilities, and yet does not lose the glamour which was never on sea or land. No artist could give more cultured notions of fairyland. In his work the vulgar glories of a pantomime are replaced by well-conceived splendour; the tawdry adjuncts of a throne-room, as represented in a theatre, are ignored. Temples and palaces of the early Renaissance, filled with graceful—perhaps a shade too suave—figures, embody all the charm of the impossible country, with none of the sordid drawbacks that are common to real life. In modern dress, as in his pictures to many of Mrs. Molesworth's stories, there is a certain unlikeness to life as we know it, which does not detract from the effect of the design; but while this is perhaps distracting in stories of contemporary life, it is a very real advantage in those of folk-lore, which have no actual date, and are therefore unafraid of anachronisms of any kind. The spirit of his work is, as it should be, intensely serious, yet the conceits which are showered upon it exactly harmonise with the mood of most of the stories that have attracted his pencil. Grimm's "Household Stories," as he pictured them, are a lasting joy. The "Bluebeard" and "Jack and the Beanstalk" toy books, the "Princess Belle Etoile," and a dozen others are nursery classics, and classics also of the other nursery where children of a larger growth take their pleasure.

ILLUSTRATION FROM "ROBINSON CRUSOE."
BY WILL PAGET.
(CASSELL AND CO.)

Without a shade of disrespect towards all the other artists represented in this special number, had it been devoted solely to Mr. Walter Crane's designs, it would have been as interesting in every respect. There is probably not a single illustrator here mentioned who would not endorse such a statement. For as a maker of children's books, no one ever attempted the task he fulfilled so gaily, and no one since has beaten him on his own ground. Even Mr. Howard Pyle, his most worthy rival, has given us no wealth of colour-prints. So that the famous toy books still retain their well-merited position as the most delightful books for the nursery and the studio, equally beloved by babies and artists.

Janet casts the Flaming Sword into the Well

ILLUSTRATION FROM "ENGLISH FAIRY TALES" BY J. D. BATTEN (DAVID NUTT)

Although a complete iconography of Mr. Walter Crane's work has not yet been made, the following list of such of his children's books as I have been able to trace may be worth printing for the benefit of those who have not access to the British Museum; where, by the way, many are not included in that section of its catalogue devoted to "Crane, Walter."

The famous series of toy books by Walter Crane include: "The Railroad A B C," "The Farmyard A B C," "Sing a Song of Sixpence," "The Waddling Frog," "The Old Courtier," "Multiplication in Verse," "Chattering Jack," "How Jessie was Lost," "Grammar in Rhyme," "Annie and Jack in London," "One, Two, Buckle my Shoe," "The Fairy Ship," "Adventures of Puffy," "This Little Pig went to Market," "King Luckieboy's Party," "Noah's Ark Alphabet," "My Mother," "The Forty Thieves," "The Three Bears," "Cinderella," "Valentine and Orson," "Puss in Boots," "Old Mother

Hubbard," "The Absurd A B C," "Little Red Riding Hood," "Jack and the Beanstalk," "Blue Beard," "Baby's Own Alphabet," "The Sleeping Beauty." All these were published at sixpence. A larger series at one shilling includes: "The Frog Prince," "Goody Two Shoes," "Beauty and the Beast," "Alphabet of Old Friends," "The Yellow Dwarf," "Aladdin," "The Hind in the Wood," and "Princess Belle Etoile." All these were published from 1873 onwards by Routledge, and printed in colours by Edmund Evans.

"SO LIGHT OF FOOT, SO LIGHT OF SPIRIT." BY CHARLES ROBINSON

ILLUSTRATION FROM "ENGLISH FAIRY TALES." BY J. D. BATTEN
(DAVID NUTT)

A small quarto series Routledge published at five shillings includes: "The Baby's Opera," "The Baby's Bouquet," "The Baby's Own Æsop." Another and larger quarto, "Flora's Feast" (1889), and "Queen Summer" (1891), were both published by Cassells, who issued also "Legends for Lionel" (1887). "Pan Pipes," an oblong folio with music was issued by Routledge. Messrs. Marcus Ward produced "Slate and Pencilvania," "Pothooks and Perseverance," "Romance of the Three Rs," "Little Queen Anne" (1885-6), Hawthorne's "A Wonder Book," first published in America, is a quarto volume with elaborate designs in colour; and "The Golden Primer" (1884), two vols., by Professor Meiklejohn (Blackwood) is, like all the above, in colour.

Of a series of stories by Mrs. Molesworth the following volumes are illustrated by Mr. Crane:—"A Christmas Posy" (1888), "Carrots" (1876), "A

Christmas Child" (1886), "Christmas-tree Land" (1884), "The Cuckoo Clock" (1877), "Four Winds Farm" (1887), "Grandmother Dear" (1878), "Herr Baby" (1881), "Little Miss Peggy" (1887), "The Rectory Children" (1889), "Rosy" (1882), "The Tapestry Room" (1879), "Tell me a Story," "Two Little Waifs," "Us" (1885), and "Children of the Castle" (1890). Earlier in date are "Stories from Memel" (1864), "Stories of Old," "Children's Sayings" (1861), two series, "Poor Match" (1861), "The Merry Heart," with eight coloured plates (Cassell); "King Gab's Story Bag" (Cassell), "Magic of Kindness" (1869), "Queen of the Tournament," "History of Poor Match," "Our Uncle's Old Home" (1872), "Sunny Days" (1871), "The Turtle Dove's Nest" (1890). Later come "The Necklace of Princess Fiorimonde" (1880), the famous edition of Grimm's "Household Stories" (1882), both published by Macmillan, and C. C. Harrison's "Folk and Fairy Tales" (1885), "The Happy Prince" (Nutt, 1888). Of these the "Grimm" and "Fiorimonde" are perhaps two of the most important illustrated books noted in these pages.

ILLUSTRATION FROM "THE WONDER CLOCK." BY HOWARD PYLE
(HARPER AND BROTHERS)

ILLUSTRATION FROM "THE WONDER CLOCK." BY HOWARD
PYLE
(HARPER AND BROTHERS)

Randolph Caldecott founded a school that still retains fresh hold of the British public. But with all respect to his most loyal disciple, Mr. Hugh Thomson, one doubts if any successor has equalled the master in the peculiar subtlety of his pictured comment upon the bare text. You have but to turn to any of his toy books to see that at times each word, almost each syllable, inspired its own picture; and that the artist not only conceived the scene which the text called into being, but each successive step before and after the reported incident itself. In "The House that Jack Built," "This is the Rat that Ate the Malt" supplies a subject for five pictures. First the owner carrying in the malt, next the rat driven away by the man, then the rat peeping up into the deserted room, next the rat studying a placard upside down inscribed "four measures of malt," and finally, the gorged animal sitting upon an empty measure. So "This is the Cat that Killed the Rat" is expanded into five pictures. The dog has four, the cat three, and the rest of the story is amplified with its secondary incidents duly sought and depicted. This literary expression is possibly the most marked characteristic of a facile and able draughtsman. He studied his subject as no one else ever studied it—he must have played with it, dreamed of it, worried it night and day, until he knew it ten times better than its author. Then he portrayed it simply and with irresistible vigour, with a fine economy of line and colour; when colour is added, it is mainly as a gay convention, and not closely imitative of nature. The sixteen toy books which bear his name are too well known to make a list of their titles necessary. A few other children's books—"What the Blackbird Said" (Routledge, 1881), "Jackanapes," "Lob-lie-by-the-Fire," "Daddy Darwin's Dovecot," all by Mrs. Ewing (S.P.C.K.), "Baron Bruno" (Macmillan), "Some of Æsop's Fables" (Macmillan), and one or two others, are of secondary importance from our point of view here.

ILLUSTRATION FROM
"THE WONDER CLOCK."
BY HOWARD PYLE
(HARPER AND BROTHERS. 1894)

ILLUSTRATION FROM "THE WONDER CLOCK." BY HOWARD PYLE. (HARPER AND BROTHERS)

It is no overt dispraise to say of Miss Kate Greenaway that few artists made so great a reputation in so small a field. Inspired by the children's books of 1820 (as a reference to a design, "Paths of Learning," reproduced on p. 9 will show), and with a curious naïvety that was even more unconcerned in its dramatic effect than were the "missal marge" pictures of the illuminators, by her simple presentation of the childishness of childhood she won all hearts. Her little people are the *beau-idéal* of nursery propriety—clean, good-tempered, happy small gentlefolk. For, though they assume peasants' garb, they never betray boorish manners. Their very abandon is only that of nice little people in play-hours, and in their wildest play the penalties that await torn knickerbockers or soiled frocks are not absent from their minds. Whether they really interested children as they delighted their elders is a moot point. The verdict of many modern children is unanimous in praise, and possibly because they represented the ideal every properly educated child is supposed to cherish. The slight taint of priggishness which occasionally is there did not reveal itself to a child's eye. Miss Greenaway's art, however, is not one to analyse but to enjoy. That she is a most careful and painstaking worker is a fact, but one that would not in itself suffice to arouse one's praise.

The absence of effort which makes her work look happy and without effort is not its least charm. Her gay yet "cultured" colour, her appreciation of green chairs and formal gardens, all came at the right time. The houses by a Norman Shaw found a Morris and a Liberty ready with furniture and fabrics, and all sorts of manufacturers devoting themselves to the production of pleasant objects, to fill them; and for its drawing-room tables Miss Greenaway produced books that were in the same key. But as the architecture and the fittings, at their best, proved to be no passing whim, but the germ of a style, so her illustration is not a trifling sport, but a very real, if small, item in the history of the evolution of picture-books. Good taste is the prominent feature of her work, and good taste, if out of fashion for a time, always returns, and is treasured by future generations, no matter whether it be in accord with the expression of the hour or distinctly archaic. Time is a very stringent critic, and much that passed as tolerably good taste when it fell in with the fashion, looks hopelessly vulgar when the tide of popularity has retreated. Miss Greenaway's work appears as refined ten years after its "boom," as it did when it was at the flood. That in itself is perhaps an evidence of its lasting power; for ten or a dozen years impart a certain shabby and worn aspect that has no flavour of the antique as a saving virtue to atone for its shortcomings.

ILLUSTRATION FROM "THE WONDER CLOCK." BY HOWARD PYLE. (HARPER AND BROTHERS)

It seems almost superfluous to give a list of the principal books by Miss Kate Greenaway, yet for the convenience of collectors the names of the most noteworthy volumes may be set down. Those with coloured plates are: "A, Apple Pie" (1886), "Alphabet" (1885), "Almanacs" (from 1882 yearly), "Birthday Book" (1880), "Book of Games" (1889), "A Day in a Child's Life" (1885), "King Pepito" (1889), "Language of Flowers" (1884), "Little Ann" (1883), "Marigold Garden" (1885), "Mavor's Spelling Book" (1885), "Mother Goose" (1886), "The Pied Piper of Hamelin" (1889), "Painting Books" (1879 and 1885), "Queen Victoria's Jubilee Garland" (1887), "Queen of the Pirate Isle" (1886), "Under the Window" (1879). Others with black-and-white illustrations include "Child of the Parsonage" (1874), "Fairy Gifts" (1875), "Seven Birthdays" (1876), "Starlight Stories" (1877), "Topo" (1878), "Dame Wiggins of Lee" (Allen, 1885), "Stories from the Eddas" (1883).

Many designs, some in colour, are to be found in volumes of *Little Folks*, *Little Wideawake*, *Every Girl's Magazine*, *Girl's Own Paper*, and elsewhere.

ILLUSTRATION FROM "CHILDREN'S SINGING GAMES" BY
WINIFRED SMITH (DAVID NUTT. 1894)

The art of Miss Greenaway is part of the legend of the æsthetic craze, and while its storks and sunflowers have faded, and some of its eccentricities are forgotten, the quaint little pictures on Christmas cards, in toy books, and elsewhere, are safely installed as items of the art product of the century. Indeed, many a popular Royal Academy picture is likely to be forgotten before the illustrations from her hand. *Bric-à-brac* they were, but more than that, for they gave infinite pleasure to thousands of children of all ages, and if they do not rise up and call her blessed, they retain a very warm memory of one who gave them so much innocent pleasure.

ILLUSTRATION FROM "UNDINE" BY HEYWOOD SUMNER
(CHAPMAN AND HALL)

ILLUSTRATION FROM "THE RED FAIRY BOOK" BY L. SPEED
(LONGMANS, GREEN AND CO. 1895)

Sir John Tenniel's illustrations, beginning as they do with "Undine" (1845), already mentioned, include others in volumes for young people that need not be quoted. But with his designs for "Alice in Wonderland" (Macmillan, 1866), and "Through the Looking Glass" (1872), we touch *the* two most notable children's books of the century. To say less would be inadequate and to say more needless. For every one knows the incomparable inventions which "Lewis Carroll" imagined and Sir John Tenniel depicted. They are veritable classics, of which, as it is too late to praise them, no more need be said.

Certain coloured picture books by J. E. Rogers were greeted with extravagant eulogy at the time they appeared "in the seventies." "Worthy to be hung at the Academy beside the best pictures of Millais or Sandys," one fatuous critic observed. Looking over their pages again, it seems strange that their very weak drawing and crude colour could have satisfied people familiar with Mr. Walter Crane's masterly work in a not dissimiliar style. "Ridicula Rediviva" and "Mores Ridiculi" (both Macmillan), were illustrations of nursery rhymes. To "The Fairy Book" (1870), a selection of old stories re-told by the author of "John Halifax," Mr. Rogers contributed many full pages in colour, and also to Mr. F. C. Burnand's "Present Pastimes of Merrie England" (1872). They are interesting as documents, but not as art; for their lack of academic knowledge is not counterbalanced by peculiar "feeling" or ingenious conceit. They are merely attempts to do again what Mr. H. S. Marks had done better previously. It seems ungrateful to condemn books that but for renewed acquaintance might have kept the glamour of the past; and yet, realising how much feeble effort has been praised since it was "only for children," it is impossible to keep silence when the truth is so evident.

ILLUSTRATION FROM "KATAWAMPUS" BY ARCHIE
MACGREGOR (DAVID NUTT)

Alfred Crowquill most probably contributed all the pictures to "Robinson Crusoe," "Blue Beard," and "Red Riding Hood" told in rhyme by F. W. N. Bayley, which have been noticed among his books of the "forties." One of the full pages, which appear to be lithographs, is clearly signed. He also illustrated the adventures of "Master Tyll Owlglass," an edition of "Baron Munchausen," "Picture Fables," "The Careless Chicken," "Funny Leaves for the Younger Branches," "Laugh and Grow Thin," and a host of other volumes. Yet the pictures in these, amusing as they are in their way, do not seem likely to attract an audience again at any future time.

E. V. B., initials which stand for the Hon. Mrs. Boyle, are found on many volumes of the past twenty-five years which have enjoyed a special reputation. Certainly her drawings, if at times showing much of the amateur, have also a curious "quality," which accounts for the very high praise they have won from critics of some standing. "The Story without an End," "Child's Play" (1858), "The New Child's Play," "The Magic Valley," "Andersen Fairy Tales" (Low, 1882), "Beauty and the Beast" (a quarto with colour-prints by Leighton Bros.), are the most important. Looking at them dispassionately now, there is yet a trace of some of the charm that provoked applause a little more than they deserve.

In British art this curious fascination exerted by the amateur is always confronting us. The work of E. V. B. has great qualities, yet any pupil of a board school would draw better. Nevertheless it pleases more than academic technique of high merit that lacks just that one quality which, for want of a better word, we call "culture." In the designs by Louisa, Marchioness of Waterford, one encounters genius with absolutely faltering technique; and many who know how rare is the slightest touch of genius, forgive the equally important mastery of material which must accompany it to produce work of lasting value.

ILLUSTRATION FROM "THE SLEEPING BEAUTY." BY R. ANNING BELL (DENT AND CO.)

Mr. H. S. Marks designed two nursery books for Messrs. Routledge, and contributed to many others, including J. W. Elliott's "National Nursery Rhymes" (Novello), whence our illustration has been taken. Two series of picture books containing mediæval figures with gold background, by J. Moyr Smith, if somewhat lacking in the qualities which appeal to children, may have played a good part in educating them to admire conventional flat treatment, with a decorative purpose that was unusual in the "seventies," when most of them appeared.

- 83 -

In later years, Miss Alice Havers in "The White Swans," and "Cape Town Dicky" (Hildesheimer), and many lady artists of less conspicuous ability, have done a quantity of graceful and elaborate pictures *of* children rather than *for* children. The art of this later period shows better drawing, better colour, better composition than had been the popular average before; but it generally lacks humour, and a certain vivacity of expression which children appreciate.

ILLUSTRATION FROM "FAIRY GIFTS." BY H. GRANVILLE FELL (DENT AND CO.)

In the "sixties" and "seventies" were many illustrators of children's books who left no great mark except on the memories of those who were young enough at the time to enjoy their work thoroughly, if not very critically. Among these may be placed William Brunton, who illustrated several of the Right Hon. G. Knatchbull-Hugessen's fairy stories, "Tales at Tea Time" for instance, and was frequent among the illustrators of Hood's Annuals. Charles H. Ross (at one time editor of *Judy*) and creator of "Ally Sloper," the British Punchinello, produced at least one memorable book for children. "Queens and Kings and other Things," a folio volume printed in gold and colour, with nonsense rhymes and pictures, almost as funny as those of Edward Lear himself. "The Boy Crusoe," and many other books of somewhat ephemeral character are his, and Routledge's "Every Boy's Magazine" contains many of his designs. Just as these pages are being corrected the news of his death is announced.

ILLUSTRATION FROM
"A BOOK OF NURSERY
SONGS AND RHYMES"
BY MARY J. NEWILL
(METHUEN AND CO. 1895)

Others, like George Du Maurier, so rarely touched the subject that they can hardly be regarded as wholly belonging to our theme. Yet "Misunderstood," by Florence Montgomery (1879), illustrated by Du Maurier, is too popular to leave unnoticed. Mr. A. W. Bayes, who has deservedly won fame in other fields, illustrated "Andersen's Tales" (Warne, 1865), probably his earliest work, as a contemporary review speaks of the admirable designs "by an artist whose name is new to us."

ILLUSTRATION FROM "THE ELF-ERRANT" BY W. E. F. BRITTEN (LAWRENCE AND BULLEN. 1895)

It is a matter for surprise and regret that Mr. Howard Pyle's illustrated books are not as well known in England as they deserve to be. And this is the more vexing when you find that any one with artistic sympathy is completely converted to be a staunch admirer of Mr. Pyle's work by a sight of "The Wonder Clock," a portly quarto, published by Harper Brothers in 1894. It seems to be the only book conceived in purely Düreresque line, which can be placed in rivalry with Mr. Walter Crane's illustrated "Grimm," and wise people will be only too delighted to admire both without attempting to compare them. Mr. Pyle is evidently influenced by Dürer—with a strong trace of Rossetti—but he carries both influences easily, and betrays a strong personality throughout all the designs. The "Merry Adventures of Robin Hood" and "Otto of the Silver Hand" are two others of about the same period, and the delightful volume collected from *Harper's Young People* for the most part, entitled "Pepper and Salt," may be placed with them. All the illustrations to these are in pure line, and have the appearance of being drawn not greatly in excess of the reproduced size. Of all these books Mr. Howard Pyle is author as well as illustrator.

Of late he has changed his manner in line, showing at times, especially in "Twilight Land" (Osgood, McIlvaine, 1896), the influence of Vierge, but even in that book the frontispiece and many other designs keep to his earlier manner.

In "The Garden behind the Moon" (issued in London by Messrs. Lawrence and Bullen) the chief drawings are entirely in wash, and yet are singularly decorative in their effect. The "Story of Jack Bannister's Fortunes" shows the artist's "colonial" style, "Men of Iron," "A Modern Aladdin," Oliver Wendell Holmes' "One-Horse Shay," are other fairly recent volumes. His illustrations have not been confined to his own stories as "In the Valley," by Harold Frederic, "Stops of Various Quills" (poems by W. D. Howells), go to prove.

ILLUSTRATION FROM "SINBAD THE SAILOR" BY WILLIAM STRANG (LAWRENCE AND BULLEN. 1896)

ILLUSTRATION FROM "ALI BABA" BY J. B. CLARK (LAWRENCE AND BULLEN. 1896)

It is strange that Mr. Heywood Sumner, who, as his notable "Fitzroy Pictures" would alone suffice to prove, is peculiarly well equipped for the illustration of children's books, has done but few, and of these none are in colour. "Cinderella" (1882), rhymes by H. S. Leigh, set to music by J. Farmer, contains very pleasant decoration by Mr. Sumner. Next comes "Sintram" (1883), a notable edition of De la Motte Fouqué's romance, followed by "Undine" (in 1885). With a book on the "Parables," by A.L.O.E., published about 1884; "The Besom Maker" (1880), a volume of country ditties with the old music, and "Jacob and the Raven," with thirty-nine illustrations (Allen, 1896), the best example of his later manner, and a book which all admirers of the more severe order of "decorative illustration" will do well to preserve, the list is complete. Whether a certain austerity of line has made publishers timid, or whether the artist has declined commissions, the fact remains that the literature of the nursery has not yet had its full share from Mr. Heywood Sumner. Luckily, if its shelves are the less full, its walls are gayer by the many Fitzroy pictures he has made so effectively, which readers of THE STUDIO have seen reproduced from time to time in these pages.

Mr. H. J. Ford's work occupies so much space in the library of a modern child, that it seems less necessary to discuss it at length here, for he is found either alone or co-operating with Mr. Jacomb Hood and Mr. Lancelot Speed, in each of the nine volumes of fairy tales and true stories (Blue, Red, Green, Yellow, Pink, and the rest), edited by Mr. Andrew Lang, and published by Longmans. More than that, at the Fine Art Society in May 1895, Mr. Ford exhibited seventy-one original drawings, chiefly those for the "Yellow Fairy Book," so that his work is not only familiar to the inmates of the nursery, but to modern critics who disdain mere printed pictures and care for nothing but autograph work. Certainly his designs have often lost much by their great reduction, for many of the originals were almost as large as four of these pages. His work is full of imagination, full of detail; perhaps at times a little overcrowded, to the extent of confusion. But children are not averse from a picture that requires much careful inspection to reveal all its story; and Mr. Ford's accessories all help to reiterate the main theme. As these eight volumes have an average of 100 pictures in each, and Mr. Ford has designed the majority, it is evident that, although his work is almost entirely confined to one series, it takes a very prominent place in current juvenile literature. That he must by this time have established his position as a prime favourite with the small people goes without saying.

ILLUSTRATION FROM "THE FLAME FLOWER." BY J. F. SULLIVAN (DENT AND CO. 1896)

Mr. Leslie Brooke has also a long catalogue of notable work in this class. For since Mr. Walter Crane ceased to illustrate the long series of Mrs. Molesworth's stories, he has carried on the record. "Sheila's Mystery," "The Carved Lions," "Mary," "My New Home," "Nurse Heathcote's Story," "The Girls and I," "The Oriel Window," and "Miss Mouse and her Boys" (all Macmillan), are the titles of these books to which he has contributed. A very charming frontispiece and title to John Oliver Hobbs' "Prince Toto," which appeared in "The Parade," must not be forgotten. The most fanciful of his designs are undoubtedly the hundred illustrations to Mr. Andrew Lang's delightful collection of "Nursery Rhymes," just published by F. Warne & Co.

These reveal a store of humour that the less boisterous fun of Mrs. Molesworth had denied him the opportunity of expressing.

Mr. C. E. Brock, whose delightful compositions, somewhat in the "Hugh Thomson" manner, embellish several volumes of Messrs. Macmillan's Cranford series, has illustrated also "The Parachute," and "English Fairy and Folk Tales," by E. S. Hartland (1893), and also supplied two pictures to that most fascinating volume prized by all lovers of children, "W. V., Her Book," by W. Canton. Perhaps "Westward Ho!" should also be included in this list, for whatever its first intentions, it has long been annexed by bolder spirits in the nursery.

A. B. Frost, by his cosmopolitan fun, "understood of all people," has probably aroused more hearty laughs by his inimitable books than even Caldecott himself. "Stuff and Nonsense," and "The Bull Calf," T. B. Aldrich's "Story of a Bad Boy," and many another volume of American origin, that is now familiar to every Briton with a sense of humour, are the most widely known. It is needless to praise the literally inimitable humour of the tragic series "Our Cat took Rat Poison." In Lewis Carroll's "Rhyme? and Reason?" (1883), Mr. Frost shared with Henry Holiday the task of illustrating a larger edition of the book first published under the title of "Phantasmagoria" (1869); he illustrated also "A Tangled Tale" (1886), by the same author, and this is perhaps the only volume of British origin of which he is sole artist. Mr. Henry Holiday was responsible for the classic pictures to "The Hunting of the Snark" by Lewis Carroll (1876).

Mr. R. Anning Bell does not appear to have illustrated many books for children. Of these, the two which introduced Mr. Dent's "Banbury Cross" series are no doubt the best known. In fact, to describe "Jack the Giant Killer" and the "Sleeping Beauty" in these pages would be an insult to "subscribers from the first." A story, "White Poppies," by May Kendall, which ran through *Sylvia's Journal*, is a little too grown-up to be included; nor can the "Heroines of the Poets," which appeared in the same place, be dragged in to augment the scanty list, any more than the "Midsummer Night's Dream" or "Keats's Poems." It is singular that the fancy of Mr. Anning Bell, which seems exactly calculated to attract a child and its parent at the same time, has not been more frequently requisitioned for this purpose. In the two "Banbury Cross" volumes there is evidence of real sympathy with the text, which is by no means as usual in pictures to fairy tales as it should be; and a delightfully harmonious sense of decoration rare in any book, and still more rare in those expressly designed for small people.

ILLUSTRATION FROM "RED APPLE AND SILVER BELLS." BY
ALICE B. WOODWARD. (BLACKIE AND SON. 1897)

The amazing number of Mr. Gordon Browne's illustrations leaves a would-be iconographer appalled. So many thousand designs—and all so good—deserve a lengthened and exhaustive eulogy. But space absolutely forbids it, and as a large number cater for older children than most of the books here noticed, on that ground one may be forgiven the inadequate notice. If an illustrator deserved to attract the attention of collectors it is surely this one, and so fertile has he been that a complete set of all his work would take no little time to get together. Here are the titles of a few jotted at random: "Bonnie Prince Charlie," "For Freedom's Cause," "St. George for England," "Orange and Green," "With Clive in India," "With Wolfe in Canada," "True to the Old Flag," "By Sheer Pluck," "Held Fast for England," "For Name and Fame," "With Lee in Virginia," "Facing Death," "Devon Boys," "Nat the Naturalist," "Bunyip Land," "The Lion of St. Mark," "Under Drake's Flag," "The Golden Magnet," "The Log of the Flying Fish," "In the King's Name,"

"Margery Merton's Girlhood," "Down the Snow Stairs," "Stories of Old Renown," "Seven Wise Scholars," "Chirp and Chatter," "Gulliver's Travels," "Robinson Crusoe," "Hetty Gray," "A Golden Age," "Muir Fenwick's Failure," "Winnie's Secret" (all so far are published by Blackie and Son). "National Nursery Rhymes," "Fairy Tales from Grimm," "Sintram, and Undine," "Sweetheart Travellers," "Five, Ten and Fifteen," "Gilly Flower," "Prince Boohoo," "A Sister's Bye-hours," "Jim," and "A Flock of Four," are all published by Gardner, Darton & Co., and "Effie," by Griffith & Farran. When one realises that not a few of these books contain a hundred illustrations, and that the list is almost entirely from two publishers' catalogues, some idea of the fecundity of Mr. Gordon Browne's output is gained. But only a vague idea, as his "Shakespeare," with hundreds of drawings and a whole host of other books, cannot be even mentioned. It is sufficient to name but one—say the example from "Robinson Crusoe" (Blackie), reproduced on page 32—to realise Mr. Gordon Browne's vivid and picturesque interpretation of fact, or "Down the Snow Stairs" (Blackie), also illustrated, with a grotesque owl-like creature, to find that in pure fantasy his exuberant imagination is no less equal to the task. In "Chirp and Chatter" (Blackie), fifty-four illustrations of animals masquerading as human show delicious humour. At times his technique appears somewhat hasty, but, as a rule, the method he adopts is as good as the composition he depicts. He is in his own way the leader of juvenile illustration of the non-Dürer school.

ILLUSTRATION FROM "KATAWAMPUS." BY ARCHIE MACGREGOR. (DAVID NUTT)

ILLUSTRATION FROM "TO TELL THE KING THE SKY IS FALLING." BY ALICE WOODWARD (BLACKIE AND SON. 1896)

Mr. Harry Furniss's coloured toy-books—"Romps"—are too well known to need description, and many another juvenile volume owes its attraction to his facile pencil. Of these, the two later "Lewis Caroll's"—"Sylvia and Bruno," and "Sylvia and Bruno, Concluded," are perhaps most important. As a curious narrative, "Travels in the Interior" (of a human body) must not be forgotten. It certainly called forth much ingenuity on the part of the artist. In "Romps," and in all his work for children, there is an irrepressible sense of movement and of exuberant vitality in his figures; but, all the same, they are more like Fred Walker's idyllic youngsters having romps than like real everyday children.

ILLUSTRATION FROM "RUSSIAN FAIRY TALES" BY C. M. GERE
(LAWRENCE AND BULLEN. 1893)

Mr. Linley Sambourne's most ingenious pen has been all too seldom employed on children's books. Indeed, one that comes first to memory, the "New Sandford and Merton" (1872), is hardly entitled to be classed among them, but the travesty of the somewhat pedantic narrative, interspersed with fairly amusing anecdotes, that Thomas Day published in 1783, is superb. No matter how familiar it may be, it is simply impossible to avoid laughing anew at the smug little Harry, the sanctimonious tutor, or the naughty Tommy, as Mr. Sambourne has realised them. The "Anecdotes of the Crocodile" and "The Presumptuous Dentist" are no less good. The way he has turned a prosaic hat-rack into an instrument of torture would alone mark Mr. Sambourne as a comic draughtsman of the highest type. Nothing he has done in political cartoons seems so likely to live as these burlesques. A little known book, "The Royal Umbrella" (1888), which contains the delightful "Cat Gardeners" here reproduced, and the very well-known edition of Charles Kingsley's "Water Babies" (1886), are two other volumes which well display his moods of less unrestrained humour. "The Real Robinson Crusoe" (1893) and Lord Brabourne's (Knatchbull-Hugessen's) "Friends and Foes of Fairyland" (1886), well-nigh exhaust the list of his efforts in this direction.

THE SINGING LESSON
No. 1. FROM THE
ORIGINAL DRAWING
BY A. NOBODY

THE SINGING LESSON
—No. 2. FROM THE
ORIGINAL DRAWING
BY A. NOBODY

ILLUSTRATION FROM "ADVENTURES IN TOY LAND" BY
ALICE B. WOODWARD (BLACKIE AND SON. 1897)

Prince of all foreign illustrators for babyland is M. Boutet de Monvel, whose works deserve an exhaustive monograph. Although comparatively few of his books are really well known in England, "Little Folks" contains a goodly number of his designs. La Fontaine's "Fables" (an English edition of which is published by the Society for Promoting Christian Knowledge) is (so far as I have discovered) the only important volume reprinted with English text. Possibly his "Jeanne d'Arc" ought not to be named among children's books, yet the exquisite drawing of its children and the unique splendour the artist has imparted to simple colour-printing, endear it to little ones no less than adults. But it would be absurd to suppose that readers of THE STUDIO do not know this masterpiece of its class, a book no artistic without. Earlier books by M. de Monvel, which show him in his most engaging mood (the mood in the illustration from "Little Folks" here reproduced), are "Vieilles Chansons et Rondes," by Ch. M. Widor, "La Civilité Puérile et Honnête," and "Chansons de France pour les Petits Français." Despite their entirely different characterisation of the child, and a much stronger grasp of the principles of decorative composition, these delightful designs are more nearly akin to those of Miss Kate Greenaway than are any others published in

Europe or America. Yet M. de Monvel is not only absolutely French in his types and costumes but in the movement and expression of his serious little people, who play with a certain demure gaiety that those who have watched French children in the Gardens of the Luxembourg or Tuileries, or a French seaside resort, know to be absolutely truthful. For the Gallic *bébé* certainly seems less "rampageous" than the English urchin. A certain daintiness of movement and timidity in the boys especially adds a grace of its own to the games of French children which is not without its peculiar charm. This is singularly well caught in M. de Monvel's delicious drawings, where naïvely symmetrical arrangement and a most admirable simplicity of colour are combined. Indeed, of all non-English artists who address the little people, he alone has the inmost secret of combining realistic drawing with sumptuous effects in conventional decoration.

ILLUSTRATION FROM "PRINCE BOOHOO" BY GORDON BROWNE (GARDNER, DARTON AND CO. 1897)

The work of the Danish illustrator, Lorenz Froelich, is almost as familiar in English as in Continental nurseries, yet his name is often absent from the title-pages of books containing his drawings. Perhaps those attributed to him formally that are most likely to be known by British readers are in "When I was a Little Girl" and "Nine Years Old" (Macmillan), but, unless memory is treacherous, one remembers toy-books in colours (published by Messrs.

Nelson and others), that were obviously from his designs. A little known French book, "Le Royaume des Gourmands," exhibits the artist in a more fanciful aspect, where he makes a far better show than in some of his ultra-pretty realistic studies. Other French volumes, "Histoire d'un Bouchée de Pain," "Lili à la Campagne," "La Journée de Mademoiselle Lili," and the "Alphabet de Mademoiselle Lili," may possibly be the original sources whence the blocks were borrowed and adapted to English text. But the veteran illustrator has done far too large a number of designs to be catalogued here. For grace and truth, and at times real mastery of his material, no notice of children's artists could abstain from placing him very high in their ranks.

Oscar Pletsch is another artist—presumably a German—whose work has been widely republished in England. In many respects it resembles that of Froelich, and is almost entirely devoted to the daily life of the inmates of the nursery, with their tiny festivals and brief tragedies. It would seem to appeal more to children than their elders, because the realistic transcript of their doings by his hand often lacks the touch of pathos, or of grown-up humour that finds favour with adults.

The mass of children's toy-books published by Messrs. Dean, Darton, Routledge, Warne, Marcus Ward, Isbister, Hildesheimer and many others cannot be considered exhaustively, if only from the fact that the names of the designers are frequently omitted. Probably Messrs. Kronheim & Co., and other colour-printers, often supplied pictures designed by their own staff. Mr. Edmund Evans, to whom is due a very large share of the success of the Crane, Caldecott, and Kate Greenaway (Routledge) books, more frequently reproduced the work of artists whose names were considered sufficiently important to be given upon the books themselves. A few others of Routledge's toy-books besides those mentioned are worth naming. Mr. H.S. Marks, R.A., designed two early numbers of their shilling series: "Nursery Rhymes" and "Nursery Songs;" and to J. D. Watson may be attributed the "Cinderella" in the same series. Other sixpenny and shilling illustrated books were by C. H. Bennett, C. W. Cope, A. W. Bayes, Julian Portch, Warwick Reynolds, F. Keyl, and Harrison Weir.

ILLUSTRATION FROM "NONSENSE" BY A. NOBODY
(GARDNER, DARTON AND CO.)

The "Greedy Jim," by Bennett, is only second to "Struwwlpeter" itself, in its lasting power to delight little ones. If out of print it deserves to be revived.

ILLUSTRATION (REDUCED) FROM "THE CHILD'S PICTORIAL." BY MRS. R. HALLWARD (S.P.C.K.)

Although Mr. William de Morgan appears to have illustrated but a single volume, "On a Pincushion," by Mary de Morgan (Seeley, 1877), yet that is so interesting that it must be noticed. Its interest is double—first in the very "decorative" quality of its pictures, which are full of "colour" and look like woodcuts more than process blocks; and next in the process itself, which was the artist's own invention. So far as I gather from Mr. de Morgan's own explanation, the drawings were made on glass coated with some yielding substance, through which a knife or graver cut the "line." Then an electro was taken. This process, it is clear, is almost exactly parallel with that of wood-cutting—*i.e.*, the "whites" are taken out, and the sweep of the tool can be guided by the worker in an absolutely untrammelled way. Those who love the qualities of a woodcut, and have not time to master the technique of wood-cutting or engraving, might do worse than experiment with Mr. de Morgan's process. A quantity of proofs of designs he executed—but never published—show that it has many possibilities worth developing.

ILLUSTRATION FROM "A, B, C" BY MRS. GASKIN (ELKIN MATHEWS)

The work of Reginald Hallward deserves to be discussed at greater length than is possible here. His most important book (printed finely in gold and colours by Edmund Evans), is "Flowers of Paradise," issued by Macmillan some years ago. The drawings for this beautiful quarto were shown at one of the early Arts and Crafts Exhibitions. Some designs, purely decorative, are interspersed among the figure subjects. "Quick March," a toy-book (Warne), is also full of the peculiar "quality" which distinguishes Mr. Hallward's work, and is less austere than certain later examples. The very notable magazine, *The Child's Pictorial*, illustrated almost entirely in colours, which the Society for Promoting Christian Knowledge published for ten years, contains work by this artist, and a great many illustrations by Mrs. Hallward, which alone would serve to impart value to a publication that has (as we have pointed out elsewhere) very many early examples by Charles Robinson, and capital work by W. J. Morgan. Mrs. Hallward's work is marked by strong Pre-Raphaelite feeling, although she does not, as a rule, select old-world themes, but depicts children of to-day. Both Mr. and Mrs. Hallward eschew the "pretty-pretty" type, and are bent on producing really "decorative" pages. So that to-day, when the ideal they so long championed has become popular, it is strange to find that their work is not better known.

"KING LOVE. A CHRISTMAS GREETING." BY H. GRANVILLE FELL

The books illustrated by past or present students of the Birmingham School will be best noticed in a group, as, notwithstanding some distinct individuality shown by many of the artists, especially in their later works, the idea that links the group together is sufficiently similar to impart to all a certain resemblance. In other words, you can nearly always pick out a "Birmingham" illustration at a glance, even if it would be impossible to confuse the work of Mr. Gaskin with that of Miss Levetus.

ILLUSTRATION FROM "THE STORY OF BLUEBEARD" BY E. SOUTHALL (LAWRENCE AND BULLEN. 1895)

Arthur Gaskin's illustrations to Andersen's "Stories and Fairy Tales" (George Allen) are beyond doubt the most important volumes in any way connected with the school. Mr. William Morris ranked them so highly that Mr. Gaskin was commissioned to design illustrations for some of the Kelmscott Press books, and Mr. Walter Crane has borne public witness to their excellence. This alone is sufficient to prove that they rise far above the average level. "Good King Wenceslas" (Cornish Bros.) is another of Mr. Gaskin's books—

his best in many ways. He it is also who illustrated and decorated Mr. Baring-Gould's "A Book of Fairy Tales" (Methuen).

Mrs. Gaskin (Georgie Cave France) is also familiar to readers of THE STUDIO. Perhaps her "A, B, C." (published by Elkin Mathews), and "Horn Book Jingles" (The Leadenhall Press), a unique book in shape and style, contain the best of her work so far.

Miss Levetus has contributed many illustrations to books. Among the best are "Turkish Fairy Tales" (Lawrence and Bullen), and "Verse Fancies" (Chapman and Hall).

"Russian Fairy Tales" (Lawrence and Bullen) is distinguished by the designs of C. M. Gere, who has done comparatively little illustration; hence the book has more than usual interest, and takes a far higher artistic rank than its title might lead one to expect.

Miss Bradley has illustrated one of Messrs. Blackie's happiest volumes this year. "Just Forty Winks" (from which one picture is reproduced here), shows that the artist has steered clear of the "Alice in Wonderland" model, which the author can hardly be said to have avoided. Miss Bradley has also illustrated the prettily decorated book of poems, "Songs for Somebody," by Dollie Radford (Nutt). The two series of "Children's Singing Games" (Nutt) are among the most pleasant volumes the Birmingham school has produced. Both are decorated by Winifred Smith, who shows considerable humour as well as ingenuity.

Among volumes illustrated, each by the members of the Birmingham school, are "A Book of Pictured Carols" (George Allen), and Mr. Baring-Gould's "Nursery Rhymes" (Methuen). Both these volumes contain some of the most representative work of Birmingham, and the latter, with its rich borders and many pictures, is a book that consistently maintains a very fine ideal, rare at any time, and perhaps never before applied to a book for the nursery. Indeed were it needful to choose a single book to represent the school, this one would stand the test of selection.

ILLUSTRATION FROM "NURSERY RHYMES" BY PAUL WOODROFFE (GEORGE ALLEN. 1897)

In Messrs. Dent's "Banbury Cross" series, the Misses Violet and Evelyn Holden illustrated "The House that Jack Built"; Sidney Heath was responsible for "Aladdin," and Mrs. H. T. Adams decorated "Tom Thumb, &c."

Mr. Laurence Housman is more than an illustrator of fairy tales; he is himself a rare creator of such fancies, and has, moreover, an almost unique power of conveying his ideas in the medium. His "Farm in Fairyland" and "A House of Joy" (both published by Kegan Paul and Co.) have often been referred to in THE STUDIO. Yet, at the risk of reiterating what nobody of taste doubts, one must place his work in this direction head and shoulders above the crowd—even the crowd of excellent illustrators—because its amazing fantasy and caprice are supported by cunning technique that makes the whole work a "picture," not merely a decoration or an interpretation of the text. As a spinner of entirely bewitching stories, that hold a child spell-bound, and can be read and re-read by adults, he is a near rival of Andersen himself.

H. Granville Fell, better known perhaps from his decorations to "The Book of Job," and certain decorated pages in the *English Illustrated Magazine*, illustrated three of Messrs. Dent's "Banbury Cross" series—"Cinderella,

&c.," "Ali Baba," and "Tom Hickathrift." His work in these is full of pleasant fancy and charming types.

A very sumptuous setting of the old fairy tale, "Beauty and the Beast," in this case entitled "Zelinda and the Monster" (Dent, 1895), with ten photogravures after paintings by the Countess of Lovelace, must not be forgotten, as its text may bring it into our present category.

Miss Rosie Pitman, in "Maurice and the Red Jar" (Macmillan), shows much elaborate effort and a distinct fantasy in design. "Undine" (Macmillan, 1897) is a still more successful achievement.

Richard Heighway is one of the "Banbury Cross" illustrators in "Blue Beard," &c. (Dent), and has also pictured Æsop's "Fables," with 300 designs (in Macmillan's Cranford series).

Mr. J. F. Sullivan—who must not be confused with his namesake—is one who has rarely illustrated works for little children, but in the famous "British Workman" series in *Fun*, in dozens of Tom Hood's "Comic Annuals," and elsewhere, has provoked as many hearty laughs from the nursery as from the drawing-room. In "The Flame Flower" (Dent) we find a side-splitting volume, illustrated with 100 drawings by the author. For this only Mr. J. F. Sullivan has plunged readers deep in debt, and when one recalls the amazing number of his delicious absurdities in the periodical literature of at least twenty years past, it seems astounding to find that the name of so entirely well-equipped a draughtsman is yet not the household word it should be.

E. J. Sullivan, with eighty illustrations to the Cranford edition of "Tom Brown's Schooldays," comes for once within our present limit.

J. D. Batten is responsible for the illustration of so many important collections of fairy tales that it is vexing not to be able to reproduce a selection of his drawings, to show the fertility of his invention and his consistent improvement in technique. The series, "Fairy Tales of the British Empire," collected and edited by Mr. Jacobs, already include five volumes—English, More English, Celtic, More Celtic, and Indian, all liberally illustrated by J. D. Batten, as are "The Book of Wonder Voyages," by J. Jacobs (Nutt), and "Fairy Tales from the Arabian Nights," edited by E. Dixon, and a second series, both published by Messrs. J. M. Dent and Co. "A Masque of Dead Florentines" (Dent) can hardly be brought into our subject.

Louis Davis has illustrated far too few children's books. His Fitzroy pictures show how delightfully he can appeal to little people, and in "Good Night Verses," by Dollie Radford (Nutt), we have forty pages of his designs that are peculiarly dainty in their quality, and tender in their poetic interpretation of child-life.

"Wymps" (Lane, 1896), with illustrations by Mrs. Percy Dearmer, has a quaint straightforwardness, of a sort that exactly wins a critic of the nursery.

J. C. Sowerby, a designer for stained glass, in "Afternoon Tea" (Warne, 1880), set a new fashion for "æsthetic" little quartos costing five or six shillings each. This was followed by "At Home" (1881), and "At Home Again" (1886, Marcus Ward), and later by "Young Maids and Old China." These, despite their popularity, display no particular invention. For the real fancy and "conceit" of the books you have to turn to their decorative borders by Thomas Crane. This artist, collaborating with Ellen Houghton, contributed two other volumes to the same series, "Abroad" (1882), and "London Town" (1883), both prime favourites of their day.

Lizzie Lawson, in many contributions for *Little Folks* and a volume in colours, "Old Proverbs" (Cassell), displayed much grace in depicting children's themes.

Nor among coloured books of the "eighties" must we overlook "Under the Mistletoe" (Griffith and Farran, 1886), and "When all is Young" (Christmas Roses, 1886); "Punch and Judy," by F. E. Weatherley, illustrated by Patty Townsend (1885); "The Parables of Our Lord," really dignified pictures compared with most of their class, by W. Morgan; "Puss in Boots," illustrated by S. Caldwell; "Pets and Playmates" (1888); "Three Fairy Princesses," illustrated by Paterson (1885); "Picture Books of the Fables of Æsop," another series of quaintly designed picture books, modelled on Struwwlpeter; "The Robbers' Cave," illustrated by A. M. Lockyer, and "Nursery Numbers" (1884), illustrated by an amateur named Bell, all these being published by Messrs. Marcus Ward and Co., who issued later, "Where Lilies Grow," a very popular volume, illustrated in the "over-pretty" style by Mrs. Stanley Berkeley. The attractive series of toy-books in colours, published in the form of a Japanese folding album, were probably designed by Percy Macquoid, and published by the same firm, who issued an oblong folio, "Herrick's Content," very pleasantly decorated by Mrs. Houghton. R. Andre was (and for all I know is still) a very prolific illustrator of children's coloured books. "The Cruise of the Walnut Shell" (Dean, 1881); "A Week Spent in a Glass Pond" (Gardner, Darton and Co.); "Grandmother's Thimble" (Warne, 1882); "Pictures and Stories" (Warne, 1882); "Up Stream" (Low, 1884); "A Lilliputian Opera" (Day, 1885); the Oakleaf Library (six shilling volumes, Warne); and Mrs. Ewing's Verse Books (six vols. S.P.C.K.) are some of the best known. T. Pym, far less well-equipped as a draughtsman, shows a certain childish naïveté in his (or was it her?) "Pictures from the Poets" (Gardner, Darton and Co.); "A, B, C" (Gardner, Darton and Co.); "Land of Little People" (Hildesheimer, 1886); "We are Seven" (1880); "Children Busy" (1881); "Snow Queen" (Gardner, Darton and Co.); "Child's Own Story Book" (Gardner, Darton and Co.).

Ida Waugh in "Holly Berries" (Griffith and Farran, 1881); "Wee Babies" (Griffith and Farran, 1882); "Baby Blossoms," "Tangles and Curls," and many other volumes mainly devoted to pictures of babies and their doings, pleased a very large audience both here and in the United States. "Dreams, Dances and Disappointments," and "The Maypole," both by Konstan and Castella, are gracefully decorated books issued by Messrs. De La Rue in 1882, who also published "The Fairies," illustrated by [H?] Allingham in 1881. Major Seccombe in "Comic Sketches from History" (Allen, 1884), and "Cinderella" (Warne, 1882), touched our theme; a large number of more or less comic books of military life and social satire hardly do so. Coloured books of which I have failed to discover copies for reference, are: A. Blanchard's "My Own Dolly" (Griffith and Farran, 1882); "Harlequin Eggs," by Civilly (Sonnenschein, 1884); "The Nodding Mandarin," by L. F. Day (Simpkin, 1883); "Cats-cradle," by C. Kendrick (Strahan, 1886); "The Kitten Pilgrims," by A. Ballantyne (Nisbet, 1887); "Ups and Downs" (1880), and "At his Mother's Knee" (1883), by M. J. Tilsey. "A Winter Nosegay" (Sonnenschein, 1881); "Pretty Peggy," by Emmet (Low, 1881); "Children's Kettledrum," by M. A. C. (Dean, 1881); "Three Wise Old Couples," by Hopkins (Cassell, 1881); "Puss in Boots," by E. K. Johnson (Warne); "Sugar and Spice and all that's Nice" (Strahan, 1881); "Fly away, Fairies," by Clarkson (Griffith and Farran, 1882); "The Tiny Lawn Tennis Club" (Dean, 1882); "Little Ben Bate," by M. Browne (Simpkin, 1882); "Nursery Night," by E. Dewane (Dean, 1882); "New Pinafore Pictures" (Dean, 1882); "Rumpelstiltskin" (De la Rue, 1882); "Baby's Debut," by J. Smith (De la Rue, 1883); "Buckets and Spades" (Dean, 1883); "Childhood" (Warne, 1883); "Dame Trot" (Chapman and Hall, 1883); "In and Out," by Ismay Thorne (Sonnenschein, 1884); "Under Mother's Wing," by Mrs. Clifford (Gardner, Darton, 1883); "Quacks" (Ward and Lock, 1883); "Little Chicks" (Griffith and Farran, 1883); "Talking Toys," "The Talking Clock," H. M. Bennett; "Four Feet by Two," by Helena Maguire; "Merry Hearts," "Cosy Corners," and "A Christmas Fairy," by Gordon Browne (all published by Nisbet).

Among many books elaborately printed by Messrs. Hildesheimer, are two illustrated by M. E. Edwards and J. C. Staples, "Told in the Twilight" (1883); and "Song of the Bells" (1884); and one by M. E. Edwards only, "Two Children"; others by Jane M. Dealy, "Sixes and Sevens" (1882), and "Little Miss Marigold" (1884); "Nursery Land," by H. J. Maguire (1888), and "Sunbeams," by E. K. Johnson and Ewart Wilson (1887).

F. D. Bedford, who illustrated and decorated "The Battle of the Frogs and Mice" (Methuen), has produced this year one of the most satisfactory books with coloured illustrations. In "Nursery Rhymes" (Methuen), the pictures, block-printed in colour by Edmund Evans, are worthy to be placed beside the best books he has produced.

Of all lady illustrators—the phrase is cumbrous, but we have no other—Miss A. B. Woodward stands apart, not only by the vigour of her work, but by its amazing humour, a quality which is certainly infrequent in the work of her sister-artists. The books she has illustrated are not very many, but all show this quality. "Banbury Cross," in Messrs. Dent's Series is among the first. In "To Tell the King the Sky is Falling" (Blackie, 1896) there is a store of delicious examples, and in "The Brownies" (Dent, 1896), the vigour of the handling is very noticeable. In "Eric, Prince of Lorlonia" (Macmillan, 1896), we have further proof that these characteristics are not mere accidents, but the result of carefully studied intention, which is also apparent in the clever designs for the covers of Messrs. Blackie's Catalogue, 1896-97. This year, in "Red Apple and Silver Bells," Miss Woodward shows marked advance. The book, with its delicious rhymes by Hamish Hendry, is one to treasure, as is also her "Adventures in Toy Land," designs marked by the *diablerie* of which she, alone of lady artists, seems to have the secret. In this the wooden, inane expression of the toys contrasts delightfully with the animate figures.

Mr. Charles Robinson is one of the youngest recruits to the army of illustrators, and yet his few years' record is both lengthy and kept at a singularly high level. In the first of his designs which attracted attention we find the half-grotesque, half-real child that he has made his own—fat, merry little people, that are bubbling over with the joy of mere existence. "Macmillan's Literary Primers" is the rather ponderous title of these booklets which cost but a few pence each, and are worth many a half-dozen high-priced nursery books. Stevenson's "Child's Garden of Verse," his first important book, won a new reputation by reason of its pictures. Then came "Æsop's Fables," in Dent's "Banbury Cross" Series. The next year saw Mr. Gabriel Setoun's book of poems, "Child World," Mrs. Meynell's "The Children," Mr. H. D. Lowry's "Make Believe," and two decorated pages in "The Parade" (Henry and Co.). The present Christmas will see several books from his hand.

"Old World Japan" (George Allen) has thirty-four, and "Legends from River and Mountain," forty-two, pictures by T. H. Robinson, which must not be forgotten. "The Giant Crab" (Nutt), and "Andersen" (Bliss, Sands), are among the best things W. Robinson has yet done.

"Nonsense," by A. Nobody, and "Some More Nonsense," by A. Nobody (Gardner, Darton & Co.), are unique instances of an unfettered humour. That their apparently naïve grotesques are from the hand of a very practised draughtsman is evident at a first glance; but as their author prefers to remain anonymous his identity must not be revealed. Specimens from the published work (which is, however, mostly in colour), and facsimiles of hitherto unpublished drawings, entitled "The Singing Lesson," kindly lent by Messrs. Gardner, Darton & Co., are here to prove how merry our anonym can be. By the way, it may be well to add that the artist in question is *not* Sir Edward Burne-Jones, whose caricatures, that are the delight of children of all ages who know them, have been so far strictly kept to members of the family circle, for whom they were produced.

ILLUSTRATION FROM "LITTLE FOLKS." BY MAURICE BOUTET DE MONVEL. (CASSELL AND CO.)

The editor of THE STUDIO, to whose selection of pictures for reproduction these pages owe their chief interest, has spared no effort to show a good working sample of the best of all classes, and in the space available has certainly omitted few of any consequence—except those so very well known, as, for instance, Tenniel's "Alice" series, and the Caldecott toy-books—which it would have been superfluous to illustrate again, especially in black and white after coloured originals.

In Mrs. Field's volume already mentioned, the author says: "It has been well observed that children do not desire, and ought not to be furnished with purely realistic portraits of themselves; the boy's heart craves a hero, and the Johnny or Frank of the realistic story-book, the little boy like himself, is not in this sense a hero." This passage, referring to the stories themselves, might be applied to their illustration with hardly less force. To idealise is the normal impulse of a child. True that it can "make believe" from the most rudimentary hints, but it is much easier to do so if something not too actual is the groundwork. Figures which delight children are never wholly symbolic, mere virtues and vices materialised as personages of the anecdote. Real nonsense such as Lear concocted, real wit such as that which sparkles from Lewis

Carroll's pages, find their parallel in the pictures which accompany each text. It is the feeble effort to be funny, the mildly punning humour of the imitators, which makes the text tedious, and one fancies the artist is also infected, for in such books the drawings very rarely rise to a high level.

ILLUSTRATION FROM "GOULD'S BOOK OF FAIRY TALES."
BY ARTHUR GASKIN. (METHUEN AND CO.)

The "pretty-pretty" school, which has been too popular, especially in anthologies of mildly entertaining rhymes, is sickly at its best, and fails to retain the interest of a child. Possibly, in pleading for imaginative art, one has forgotten that everywhere is Wonderland to a child, who would be no more astonished to find a real elephant dropping in to tea, or a real miniature railway across the lawn, than in finding a toy elephant or a toy engine awaiting him. Children are so accustomed to novelty that they do not realise the abnormal; nor do they always crave for unreality. As coaches and horses were the delight of youngsters a century ago, so are trains and steamboats to-day. Given a pile of books and an empty floor space, their imagination needs no mechanical models of real locomotives; or, to be more correct, they enjoy the make-believe with quite as great a zest. Hence, perhaps, in praising conscious art for children's literature, one is unwittingly pleasing older tastes; indeed, it is not inconceivable that the "prig" which lurks in most of us may be nurtured by too refined diet. Whether a child brought up wholly on the æsthetic toy-book would realise the greatness of Rembrandt's etchings or other masterpieces of realistic art more easily than one who had only known the current pictures of cheap magazines, is not a question to be decided off-hand. To foster an artificial taste is not wholly unattended with danger; but if humour be present, as it is in the works of the best artists for the nursery, then all fear vanishes; good wholesome laughter is the deadliest bane to the prig-microbe, and will leave no infant lisping of the preciousness of Cimabue, or the wonder of Sandro Botticelli, as certain children were reported to do in the brief days when the æsthete walked his faded way among us. That modern children's books will—some of them at least—take an honourable place in an iconography of nineteenth-century art, many of the illustrations here reproduced are in themselves sufficient to prove.

ILLUSTRATION FROM "LULLABY LAND" BY CHARLES ROBINSON. (JOHN LANE. 1897)

After so many pages devoted to the subject, it might seem as if the mass of material should have revealed very clearly what is the ideal illustration for children. But "children" is a collective term, ranging from the tastes of the baby to the precocious youngsters who dip into Mudie books on the sly, and hold conversations thereon which astonish their elders when by chance they get wind of the fact. Perhaps the belief that children can be educated by the eye is more plausible than well supported. In any case, it is good that the illustration should be well drawn, well coloured; given that, whether it be realistically imitative or wholly fantastic is quite a secondary matter. As we have had pointed out to us, the child is not best pleased by mere portraits of himself; he prefers idealised children, whether naughtier and more adventurous, or absolute heroes of romance. And here a strange fact appears, that as a rule what pleases the boy pleases the girl also; but that boys look down with scorn on "girls' books." Any one who has had to do with children knows how eagerly little sisters pounce upon books owned by their brothers. Now, as a rule, books for girls are confined to stories of good girls, pictures of good girls, and mildly exciting domestic incidents, comic or tragic. The child may be half angel; he is undoubtedly half savage; a Pagan indifference to other people's pain, and grim joy in other people's accidents, bear witness to that fact. Tender-hearted parents fear lest some pictures should terrify the little ones; the few that do are those which the child himself discovers in some extraordinary way to be fetishes. He hates them, yet is fascinated by them. I remember myself being so appalled by a picture that is still keenly remembered. It fascinated me, and yet was a thing of which the mere memory made one shudder in the dark—the said picture representing a benevolent negro with Eva on his lap, from "Uncle Tom's Cabin," a blameless Sunday-school inspired story. The horrors of an early folio of Foxe's "Martyrs," of a grisly "Bunyan," with terrific pictures of Apollyon; even a still more grim series by H. C. Selous, issued by the Art Union, if memory may be trusted, were merely exciting; it was the mild and amiable representation of "Uncle Tom" that I felt to be the very incarnation of all things evil. This personal incident is quoted only to show how impossible it is for the average adult to foretell what will frighten or what will delight a child. For children are singularly reticent concerning the "bogeys" of their own creating, yet, like many fanatics, it is these which they really most fear.

ILLUSTRATION FROM "MAKE BELIEVE." BY CHARLES ROBINSON (JOHN LANE. 1896)

ILLUSTRATION FROM "JUST FORTY WINKS" BY GERTRUDE M. BRADLEY (BLACKIE AND SON. 1897)

Certainly it is possible that over-conscious art is too popular to-day. The illustrator when he is at work often thinks more of the art critic who may review his book than the readers who are to enjoy it. Purely conventional groups of figures, whether set in a landscape, or against a decorative background, as a rule fail to retain a child's interest. He wants invention and detail, plenty of incident, melodrama rather than suppressed emotion. Something moving, active, and suggestive pleases him most, something about which a story can be woven not so complex that his sense is puzzled to explain why things are as the artist drew them. It is good to educate children unconsciously, but if we are too careful that all pictures should be devoted to raising their standard of taste, it is possible that we may soon come back to the Miss Pinkerton ideal of amusement blended with instruction. Hence one doubts if the "ultra-precious" school really pleases the child; and if he refuse the jam the powder is obviously refused also. One who makes pictures for children, like one who writes them stories, should have the knack of entertaining them without any appearance of condescension in so doing. They will accept any detail that is related to the incident, but are keenly alive to discrepancies of detail or action that clash with the narrative. As they do not demand fine drawing, so the artist must be careful to offer them very much more than academic accomplishment. Indeed, he (or she) must be in sympathy with childhood, and able to project his vision back to its point of view. And this is just a mood in accord with the feeling of our own time, when men distrust each other and themselves, and keep few ideals free from doubt, except the reverence for the sanctity of childhood. Those who have forsaken beliefs hallowed by centuries, and are the most cynical and worldly-minded, yet often keep faith in one lost Atalantis—the domain of their own childhood and those who still dwell in the happy isle. To have given a happy hour to one of the least of these is peculiarly gratifying to many tired people to-day, those surfeited with success no less than those weary of failure. And such labour is of love all compact; for children are grudging in their praise, and seldom trouble to inquire who wrote their stories or painted their pictures. Consequently those who work for them win neither much gold nor great fame; but they have a most enthusiastic audience all the same. Yet when we remember that the veriest daubs and atrocious drawings are often welcomed as heartily, one is driven to believe that after all the bored people who turn to amuse the children, like others who turn to elevate the masses, are really, if unconsciously, amusing if not elevating themselves. If children's books please older people—and that they do so is unquestionable—it would be well to acknowledge it boldly, and to share the pleasure with the nursery; not to take it surreptitiously under the pretence of raising the taste of little people. Why should not grown-up people avow their pleasure in children's books if they feel it?

THE SPOTTED MIMILUS. ILLUSTRATION FROM "KING LONGBEARD." BY CHARLES ROBINSON (JOHN LANE. 1897)

ILLUSTRATION FROM "THE MAKING OF MATTHIAS" BY
LUCY KEMP-WELCH. (JOHN LANE. 1897)

If a collector in search of a new hobby wishes to start on a quest full of disappointment, yet also full of lucky possibilities, illustrated books for children would give him an exciting theme. The rare volume he hunted for in vain at the British Museum and South Kensington, for which he scanned the shelves of every second-hand bookseller within reach, may meet his eye in a twopenny box, just as he has despaired of ever seeing, much less procuring, a copy. At least twice during the preparation of this number I have enjoyed that particular experience, and have no reason to suppose it was very abnormal. To make a fine library of these things may be difficult, but it is not a predestined failure. Caxtons and Wynkyn de Wordes seem less scarce than some of these early nursery books. Yet, as we know, the former have been the quest of collectors for years, and so are probably nearly all sifted out of the great rubbish-heaps of dealers; the latter have not been in great demand, and may be unearthed in odd corners of country shops and all sorts of likely and unlikely places. Therefore, as a hobby, it offers an exciting quest with almost certain success in the end; in short, it offers the ideal conditions for collecting as a pastime, provided you can muster sufficient interest in the subject to become absorbed in its pursuit. So large is it that, even to limit one's quest to books with coloured pictures would yet require a good many years' hunting to secure a decent "bag." Another tempting point is that prices at present are mostly nominal, not because the quarry is plentiful, but because the demand is not recognised by the general bookseller. Of course, books in good condition, with unannotated pages, are rare; and some series—Felix Summerley's, for example—which owe their chief interest to the "get-up" of the volume considered as a whole, would be scarce worth possessing if "rebound" or deprived of their covers. Still, always provided the game attracts him, the hobby-horseman has fair chances, and is inspired by motives hardly less noble than those which distinguish the pursuit of bookplates (*ex libris*), postage-stamps and other objects which have attracted men to devote not only their leisure and their spare cash, but often their whole energy and nearly all their resources. Societies, with all the pomp of officials, and members proudly arranging detached letters of the alphabet after their names, exist for discussing hobbies not more important. Speaking as an interested but not infatuated collector, it seems as if the mere gathering together of rarities of this sort would soon become as tedious as the amassing of dull armorial *ex libris*, or sorting infinitely subtle varieties of postage-stamps. But seeing the intense passion such things arouse in their devotees, the fact that among children's books there are not a few of real intrinsic interest, ought not to make the hobby less attractive; except that, speaking generally, your true collector seems to despise every quality except rarity (which implies market value ultimately, if for the moment there are not enough rival collectors to have started a "boom" in prices). Yet all these "snappers up of unconsidered trifles" help to gather together material which may prove in time to be not without value to the social historian or the student interested in the progress of printing and the art of illustration; but it would be a pity to confuse ephemeral "curios" with lasting works of fine art, and the ardour of collecting need not blind one to the fact that the former are greatly in excess of the latter.

ILLUSTRATION FROM "MISS MOUSE AND HER BOYS." BY L. LESLIE BROOKE. (MACMILLAN AND CO. 1897)

The special full-page illustrations which appear in this number must not be left without a word of comment. In place of re-issuing facsimiles of actual illustrations from coloured books of the past which would probably have been familiar to many readers, drawings by artists who are mentioned elsewhere in this Christmas Number have been specially designed to carry out the spirit of the theme. For Christmas is pre-eminently the time for children's books. Mr. Robert Halls' painting of a baby, here called "The Heir to Fairyland"—the critic for whom all this vast amount of effort is annually expended—is seen still in the early or destructive stage, a curious foreshadowing of his attitude in a later development should he be led from the paths of Philistia to the bye-ways of art criticism. The portrait miniatures of child-life by Mr. Robert Halls, if not so well known as they deserve, cannot be unfamiliar to readers of THE STUDIO, since many of his best works have been exhibited at the Academy and elsewhere.

The lithograph by Mr. R. Anning Bell, "In Nooks with Books," represents a second stage of the juvenile critic when appreciation in a very acute form has set in, and picture-books are no longer regarded as toys to destroy, but treasures to be enjoyed snugly with a delight in their possession.

ILLUSTRATION FROM "BABY'S LAYS" BY E. CALVERT (ELKIN MATHEWS. 1897)

Mr. Granville Fell, with "King Love, a Christmas Greeting," turns back to the memory of the birthday whose celebration provokes the gifts which so often take the form of illustrated books, for Christmas is to Britons more and more the children's festival. The conviviality of the Dickens' period may linger here and there; but to adults generally Christmas is only a vicarious pleasure, for most households devote the day entirely to pleasing the little ones who have annexed it as their own special holiday.

The dainty water-colour by Mr. Charles Robinson, and the charming drawing in line by M. Boutet de Monvel, call for no comment. Collectors will be glad to possess such excellent facsimiles of work by two illustrators conspicuous for their work in this field. The figure by Mr. Robinson, "So Light of Foot, so Light of Spirit," is extremely typical of the personal style he has adopted from the first. Studies by M. de Monvel have appeared before in THE STUDIO, so that it would be merely reiterating the obvious to call attention to the exquisite truth of character which he obtains with rare artistry.

G. W.

The Editor's best thanks are due to all those publishers who have so kindly and readily come forward with their assistance in the compilation of "Children's Books and their Illustrators." Owing to exigences of space reference to several important new books has necessarily been postponed.

ILLUSTRATION FROM "NATIONAL RHYMES." BY GORDON BROWNE (GARDNER, DARTON AND CO. 1897)